The Intern Blues

The Timeless Classic About the Making of a Doctor

ROBERT MARION, M.D.

HARPER

NEW YORK • LONDON • TORONTO • SYDNEY

First Perennial edition published 2001.

Designed by Jeannette Jacobs

Library of Congress Cataloging-in-Publication Data

Marion, Robert.
    The intern blues : the timeless classic about the making of a doctor / Robert Marion.—[2nd ed.].
      p. cm.
    ISBN 0-06-093709-2 (alk. paper)
    1. Interns (Medicine)—United States—Biography. I. Title.

R153.M37 2001
610'.71'173—dc21
[B]

                                                            2001021502

17 18 19   WB/RRD   20 19 18

About the Author

Robert Marion, M.D., a professor of pediatrics and obstetrics and gynecology at the Albert Einstein College of Medicine in the Bronx, New York, is the director of clinical genetics at both the Montefiore Medical Center in the Bronx and Blythedale Children's Hospital, Valhalla, New York. He is the author of six published books, including *The Intern Blues* and *Learning to Play God: The Coming of Age of a Young Doctor.* He lives with his family in Westchester County, New York.

To my parents, Ann and Sam Marion,
who sweated and slaved to put me through medical school
so that I could sweat and slave as an intern

and to the pediatric house staff who
compose our program, without whom this book
would not have been possible

AUTHOR'S NOTE

Imagine walking into a room occupied by thirty-five tiny and very sick premature newborn babies. For the next twelve hours their well-being, their very survival is your responsibility. That was how my internship year began. It was definitely one of the most terrifying days of my life.

I somehow survived that year in spite of the fact that I and the other interns in my group received negligible guidance from the senior people in the program. I was overworked, overtired, lonely and insecure, often depressed, and conflicted by my own responsibilities, whether admitting an infant with a dangerously high fever or coping with the psychological and physical stresses of dealing with AIDS patients.

From July 1985 through June 1986, I worked with three interns: Amy, Andy, and Mark. Together, we've collected very personal notes—often recorded after a grueling night on call—about what it's really like, day by day, to spend a year as an intern, meeting too often with frustrations and not enough encouragement.

The Intern Blues is the product of those notes.

ROBERT MARION, M.D.

Monday, January 6, 2001

"Tell me about your worst night on call."

Without hesitation, Emily, a senior resident who was spending January doing an elective with me in Medical Genetics, responded to my request. "It's hard to single out an absolute worst night. I've had a few that would make the finals. They all happened during the winter of my internship. Winter is terrible for interns, you never see any sunlight; you get to the hospital when it's still dark, and you leave at night when it's dark again. You're never outside when it's sunny, and that's bad. But probably my single worst night was a Friday in February. I was working on the Infants' Unit at Mount Scopus, and that morning, a ten-month-old baby came in who we thought might have a septic hip **[an infection of the hip joint; a medical emergency because if left untreated, the infection can destroy the joint, leading to long-term disability]**. The intern who'd been on call Thursday night admitted her, and by the time we finished rounds in the morning, the orthopedic surgeons had whisked the child off to tap her hip before any of us had a chance to see her or meet the family. The child came back to the floor from the OR in the afternoon. About four o'clock, I got a call from the pediatric radiologist who said, 'I just reviewed the X-rays and I see multiple fractures of the hip. They make me suspicious of child abuse.' He told me to bring the kid down right away for a skeletal survey to see if there were any other signs of fractures. So I brought the kid down to the radiology department. The films were taken, and the radiologist saw what he thought were three old, healed fractures of other bones. They were subtle; he said he thought another radiologist might not have read them the same way, but he felt strongly enough about it that he thought a report needed to be made.

"So I took the baby back upstairs, and as the intern on call that night, it became my responsibility to make the phone call to Child Protective Services. I had to tell the family what was going on. That was horrible. They were Spanish-speaking, and my Spanish is not so good. Plus, since only one parent is allowed to stay in the hospital with the child, the parents were split: the father was home, and the mother was in the baby's room. As best I could, I told the mother that we'd found evidence of new and old fractures on the X rays and that we suspected child abuse . . . she was devastated, horribly devastated. She started crying, and I didn't know what to say to her to calm her down.

"Then the investigators from Child Protective Services came to speak with the mother. By this point, it was about midnight. They talked to the mother, and then they interviewed me; they wanted me to be the expert witness. They asked me questions about the radiologic findings. What did these fractures mean? How could they have happened? Is there any way they could have occurred other than through abuse? I had no idea how to answer them. And this poor mother was asking me questions, too, questions she deserved answers to. But I didn't have any answers for her, and it being the middle of the night by this point, there was nobody around who could give her those answers. It was horrible.

"Then the detectives came. First they interviewed the mother. She was horrified; she denied that she or the baby's father had done anything wrong, and I believed her. . . . I believed that she didn't have anything to do with it. The detectives asked me the same questions the Child Protective Services investigator had asked. I don't know. . . . This was the first time I'd ever dealt with anything like this, the first time I'd ever had to handle any of the legal aspects of my job. My approach to the same situation is very different today, after two years of experience dealing with these kinds of problems. But back then, I didn't know. . . . I thought, 'Here I am, the pediatrician in charge of the case. I'm supposed to know the answers to all these questions the detectives are asking me.' They were hounding me, writing down every word I said. I figured that since I'd called them and they'd come in at three o'clock in the morning, that I should have at least been able to tell them for sure whether this child had been abused or not. It was horrible, horrible for me, horrible for the mother, horrible for the family, horrible for the kid, horrible for everyone involved.

"Meanwhile, while all this was going on, it was a typical busy winter night on Infants'. By the time the detectives came, we had three or four more admissions to work up, and a few IVs had fallen out and needed to be replaced. There comes a point in the night when you're working and working and you finally realize that you're not even going to get fifteen minutes to lie down. You're not going to have time to eat, you're not going to have time to go to the bathroom. I had all this work to do, all these thoughts about the baby with the fractures going through my mind, I was exhausted, and I was having trouble concentrating. At about five thirty, I was talking to the mother of one of the new admissions, trying my hardest to stay awake and take the history. I was definitely half asleep while I was talking to her, droopy eyed, and after asking a question, this mother turned to me and said, 'Doctor, I think you already asked me that.' The fact that I had lost track of what I was doing kind of jolted me out of my sleepiness. I got really worried that I was going to screw up and make a major mistake, and so I willed myself to pay attention. Then, I walked out of the room and saw the intern from one of the other floors. She had come down to see how I was. Just seeing her face and realizing she was also awake at that hour brightened the night enough to help me get through it."

"So you got through it," I said.

"Yeah, I got through it. I got through all those horrible nights on call as an intern. It was bad."

"What happened to the baby with the fractured hip?" I asked. "Did they ever figure out how she had broken those bones?"

"Not that I know of," Emily replied. "The mother continued to deny that she'd done anything wrong. The father also denied hurting her. The investigators blamed one, then the other, then the first one again. They finally sent the baby to a foster home pending completion of the legal case. I had to appear in court. That was another horrible experience, one I haven't thought about in a while. But I don't know what ultimately happened. I never got follow-up."

As I write this, it's been over fifteen years since Andy Baron, Mark Greenberg, and Amy Horowitz, the three frightened, inexperienced medical-school graduates whose audiodiaries serve as the basis of *The Intern Blues,* and I first talked about this project. Since that morning in late June 1985, when we sat sprawled on the lawn outside the home of Peter Anderson, chairman of the Department of Pediatrics at the Albert Schweitzer School of Medicine, a lot has changed in our residency program. When Andy, Mark, and Amy began their residency, our training program was huge. Our more than 100 house officers staffed seven general pediatric inpatient wards, three neonatal intensive care units and well-baby nurseries, two pediatric intensive care units, three emergency rooms, and about six outpatient clinics. These many activities were distributed among four major teaching hospitals arrayed on two separate campuses in the Bronx. In trying to provide care for the sick infants and children who were admitted to these inpatient units, as well as for the more healthy ones who visited the ambulatory care clinics, our interns were unbelievably stressed. They were on call every third night, usually spending those nights awake in the hospital, often caring for critically ill, sometimes dying patients; having to deal with frightened and angry parents, lab technicians who didn't give a damn, intimidating and demanding nurses, and abusive attending physicians; rarely even having a glimmer of a chance of getting some sleep. They spent their days traveling from one campus to the other, providing care to their inpatient charges, then signing out and heading off to one of the other sites where a full panel of clinic patients awaited them. Since they spent most of their hours outside the hospital trying to catch up on sleep, they had little or no time for a social life, often going weeks without speaking with friends, family, or, in fact, anyone who worked outside the hospital.

Between 1985 and the late 1990s, a series of events began to pare away at our program, both internally and externally. Government subsidies for the training of young doctors shrank. Because of a presumed "doctor glut," hospitals were offered incentives to train fewer residents. A major internecine struggle between Mt. Scopus Medical Center, the major teaching hospital of the Albert Schweitzer School of Medicine, and New York City's Health and Hospitals Corporation caused our program to withdraw its trainees from our two municipal hospital affiliates (Jonas Bronck and West Bronx Medical Centers). As a result, by the time Emily began her internship in July 1998, the program had shrunk dramatically. Emily became part of a residency force of only sixty-five, less than two-thirds the number with whom Andy, Mark, and Amy had worked. She and her colleagues spent most of their training time at Mt. Scopus Medical Center, venturing out to University Hospital only for rotations in the Neonatal Intensive Care Unit. They cared for children on only three general pediatric wards, in one NICU, one well-baby nursery, one pediatric intensive care unit, one emergency room, and two outpatient general pediatric clinics.

Although the paring of the size of our program had an effect on our trainees, no single event in the lives of interns and residents had as much of an impact as the adoption of the Bell Commission Regulations. In July 1989, New York became the first state to institute laws regulating the working conditions under which young doctors could be trained. Back in the days Andy, Mark, and Amy were interns, there were no regulations governing the conditions under which house officers worked. Interns and residents typically worked 110 hours per week, spending 36 or more hours in the hospital during a single on-call shift, working 18 to 24 hours in the emergency room without relief. The new regulations changed this outrageous situation: They limited the number of hours a house officer could work to 80 per week, with no more than 24 consecutive hours as part of a single shift, and no more than 12 consecutive hours working in an emergency-room setting.

As a result, by the time Emily began her training, life had become much different for interns. Emily was on-call every fourth night instead of every third. Instead of having to hang around until five or six (or seven) P.M. after spending another sleepless night in the hospital, as had been the custom back in the days prior to the introduction of the regulations, she was encouraged to leave no later than noon on her post-call days. No longer needing to spend precious minutes commuting from one campus to the other to fulfill her patient-care responsibilities, she was able to get out of the hospital earlier on non-call days, too, contributing to her ability to have some semblance of a social life. Yes, thanks to the Bell Commission Regulations, life for house officers should have improved dramatically.

The key phrase there is "should have." "The fact is," Emily said later in our conversation, "residency was actually pretty good. But internship was horrible. Even though I was on every fourth night instead of every third, it was still brutal. By the time February rolled around, I was chronically over-tired. During the first few months of the year, I was so nervous and scared I was going to accidentally screw up and kill someone, I had trouble sleeping even during the nights I wasn't on-call. And seeing kids so sick, and watching some of them die, and dealing with their parents and families—nothing in life prepared me for that. By the time I became a resident, it wasn't as bad. I felt more sure of myself; I knew what to do for my patients, how to handle most situations. But internship? I'm glad I never have to go through that year again."

So, although the events documented in *The Intern Blues* occurred more than fifteen years ago, the message delivered by the three young doctors-in-training is still relevant. Yes, it's true that some of the events that occurred may be dated, and some of the methods used to train house officers are no longer in use. But there are things about internship that will never change. As long as medical school graduates are plunged into a setting in which they have to care for sick and dying patients over long hours with very little support from those around them, the experiences of Andy, Mark, and Amy will continue to ring true.

Since the original publication of *The Intern Blues* in 1989, I've had the opportunity to speak with a lot of people who've read the book. In addition to noting similarities between the experiences of one of or more of the interns whose lives were chronicled and their own experiences, many former residents have noted that the book was most valuable to them as a way of educating their loved ones. "My parents never understood what I was going through," many have commented. This was the case with me, as well. When I was a resident, my father used to ask me, "When you're on-call at night, what does that mean? You go home and wait for them to call if they need you?" After reading the book, he finally understood.

I think this function, *The Intern Blues* as an educational tool for non-doctors, is also still valid today. In fact, I think reading the book before beginning one's internship may be harmful to the future intern's health; rather, the book should be read by the parents, other family members, and loved ones of interns prior to the start of their training years. That's the only way they'll truly understand what hell their loved ones' life is likely to become.

PREFACE

All of the events described within this book actually occurred. Not all of them, however, involved the intern to whom they have been assigned here. In order to provide the doctors, patients, and staff with anonymity, some of the occurrences, patient contacts, and reactions have been altered or switched. As a result, some of the characterizations that emerge represent composites rather than actual portraits.

Additionally, the names of the hospitals, physicians, staff members, and patients have been changed. To render the interns even less identifiable, their physical characteristics have been appreciably altered. In spite of these changes, however, this is a work of nonfiction; the events and experiences described are all true.

This book would not have been possible without the cooperation of a large number of people. I'd like to take this opportunity to thank the faculty and administration of our pediatric program, the administrations of both the hospitals through which our interns and residents rotate and the medical school affiliated with those hospitals, the house staff who make up our program, and especially the interns who allowed me to just about live inside their heads during that very difficult year.

Finally, I'd especially like to thank the following people: my wife, Beth, and my children, Isadora, Davida, and Jonah, for putting up with the long hours I spent playing with my computer rather than playing with them; Pamela Altschul, editorial assistant at William Morrow, for her help and sharp

insights; Diana Finch, my literary agent, for her encouragement; and Adrian Zackheim, senior editor at William Morrow and self-proclaimed "medical junkie," who has been
there to guide me through every step in the writing of this
book.

The stretchers arrived at the emergency room about fifteen minutes after I started my shift. I had barely had enough time to say hello to the residents and nurses on call when suddenly, out of the clear blue, three critically ill brothers were being wheeled into the trauma area.

We all immediately went running to the back to meet them. One of the Emergency Medical Service workers yelled out an abbreviated version of their story: "It's an apartment fire. The FD [**Fire Department**] pulled them out of the bedroom. We loaded them onto the stretchers and transported." He further explained that the boys' mother was at that moment being wheeled into the adult emergency room. She was near death.

Apartment fires were unusual in May; they're usually winter events, when everyone's using space heaters to try to keep warm. But unusual or not, we swung into action. With very little discussion, we triaged them: The senior resident took the eight-year-old, who was semiconscious. The junior resident began to work on the six-year-old, who was the best off of the three: His vital signs were stable and he was awake enough to answer questions. And the two interns and I moved straight toward the ten-year-old, who was comatose; he wasn't breathing on his own, and his fingers and lips were beginning to turn purple. We knew we had to act fast.

By reflex, Amy, one of the interns, grabbed the black rubber ambubag and began to force oxygen into the boy's lungs while I went about gathering the supplies needed for intubation. I got a pediatric laryngoscope [**a light source with a metal blade at its end, designed to push away the tongue**

and illuminate the back of the throat] and an endotracheal tube from the code cart. Meanwhile, Andy, the other intern, after listening with his stethoscope, had determined that the boy's heart wasn't beating. Without a word, he immediately began pumping the chest about a hundred times a minute. At that point I heard the announcement over the loudspeaker: "Attention, attention: CAC in the pediatric emergency room. Attention attention: CAC in the pediatric emergency room." I was relieved to hear it: It meant that help was on the way.

After Amy had bagged the kid for maybe a minute, I nudged her away and got ready to do the intubation. I concentrated all my efforts on the back of that boy's throat. Holding the laryngoscope in my left hand, I placed the instrument into the patient's mouth and shifted it around until I could clearly see the vocal cords. Then I began to push the endotracheal tube through those cords. At first the tube slipped backward, falling down into the esophagus. I repositioned it and tried again. This time the tube slipped right through the cords and slid down into the trachea.

I was sure it was in, so I began to call for a piece of tape, but before I could get the words out, a healthy piece of the stuff was being dangled before my eyes. Anticipating my need, Amy had torn a supply and now all that needed to be done was to apply one end to the skin of the boy's upper lip and wrap the other end around the tube so it would remain steadily in place. It took me about a minute to secure the tube, and when I finished, I hooked up the ambubag and began to force oxygen directly into the boy's lungs.

As I compressed the ambubag, I began to take stock of the situation. The trauma area was now packed with medical personnel who, thanks to the loudspeaker announcement, had come running from all corners of the hospital. It was then that I realized that for the first time all year, Amy Horowitz, Andy Baron, and Mark Greenberg, who had answered the call for help, were all working together on a single patient.

Amy grabbed the ambubag and relieved me. Andy was continuing the chest compressions, while Mark was working on getting an IV into the boy's arm. He succeeded on the first try and was simultaneously hooking up the IV drip and asking one of the nurses for a shot of epinephrine, a drug that hopefully would help start the boy's heart beating again. Meanwhile, I began to attach the leads from an electrocardiograph machine to the boy's wrists and ankles, in an attempt to monitor the activity of his heart better. All this was being carried out without a word of direction from me. Each of us knew what had to be done and were doing it without any prompting.

It took nearly fifteen minutes to get that kid's heart started again, but after Mark had pushed in the second round of medications, electrical activity began to appear on the cardiograph paper. "We've got complexes," I said when I saw them. "It looks like a normal rhythm." The interns breathed a sigh of relief when they heard my words. Now comes Miller Time!

In another minute, the boy began breathing on his own. He was reasonably stable now, so we pulled back and took stock of what needed to be done. Amy volunteered to take charge of the boy until the intern from the ICU [**intensive-care unit**] upstairs came down to get him. His brothers, now also stable, and our patient went up to Jonas Bronck's pediatric ICU about a half hour later. After prolonged hospital stays, they each recovered and were discharged home. Their mother, however, wasn't as fortunate. She never regained consciousness and died later that night in the adult ICU.

Watching those three interns working together on that boy in such perfect harmony, with such confidence in their judgment and their technical ability, it was hard for me to believe that only eleven months earlier they had begun this internship. It seemed incredible that they were the same people who, when I had talked with them out on the lawn of Peter Anderson's house in Westchester County at orientation, had seemed so tense and uncertain and downright scared to death.

I had met Amy, Mark, and Andy at that orientation retreat at the house of the chairman of the Department of Pediatrics on June 26, 1985. All around us on the lawn, the exact same scene was being played out: Stretched out on the grass were groups of three or four new interns, each looking well rested and tanned from their month of vacation and each as tense as a turkey around Thanksgiving because of the year of torture that loomed ahead. Sitting with each group of interns was an attending physician, one of the senior doctors affiliated with the pediatric program, who would serve as teacher, mentor, and at times taskmaster to the new interns. We attendings were trying our best to convince these guys that the next twelve months weren't going to be as bad as they had been led to believe. In other words, we were lying through our teeth.

Over the past seven years, it had become traditional in the Albert Schweitzer School of Medicine's pediatric residency training program that the internship year begin with this orientation retreat. Regardless of what the day accomplishes, it's a nice idea, an opportunity for the thirty-five new interns to get to know each other in a relaxed atmosphere, to make friends with the people with whom they'll be spending every day and every third night over the next twelve months. The retreat also gives the interns a chance to meet the chief residents, the four physicians who are directly in charge of them, the people they'd have to turn to in times of crisis.

My first meeting with Amy, Mark, and Andy started out pretty disappointingly. I'd led small groups at these retreats for the past three years, and this one was definitely the hardest to get off the ground. The idea was to get the interns talking about their concerns so that they'd discover these concerns weren't unique, that the same fears were shared by each of their classmates. But for that to happen the interns had to talk, and so far they were keeping their mouths tightly shut.

I decided to cut through the small talk and take a more direct approach. "Look," I began, "I know you guys must be scared to death. You're so nervous, I'm getting jumpy just sitting here. What are you so worried about?"

There was silence again for what seemed like hours, but it was probably no more than a minute. I was thinking I'd have to come up with some other tactic when Andy Baron suddenly spoke up. "I'll tell you what I'm worried about," he said just loud enough to be heard. "I'm worried I don't know enough."

"Don't know enough about what?" I immediately asked, overjoyed that somebody had actually said something.

Andy thought for a few seconds. "I'm worried that I'm going to get out there on the wards and be expected to know certain things that I just don't know. I don't know the kinds of things doctors are supposed to know."

"What are doctors supposed to know?" I asked.

"They're supposed to know everything," Andy replied without hesitation. "They're supposed to know what to do in an emergency; they're supposed to know what's going wrong when it goes wrong and what to do to make it better. I don't know any of those things. I never had to know anything that important when I was a medical student."

"Doctors are also supposed to be able to do things like start IVs and do spinal taps," Mark Greenberg said next. "I don't know about you guys, but if I were to go into a hospital today and do a spinal tap on a baby, I could be charged with assault with a deadly weapon. I'm not sure, but I don't think a criminal record is exactly what we're trying to accomplish here."

"So you're worried that you don't know enough and that even if you did know enough, you couldn't do anything to help the patients because you don't have the technical skills," I said. "Is that about right?" The three of them nodded yes. I wrote this down on a piece of paper. As group leader, I was supposed to act as a kind of anxiety scorekeeper.

"Look," I began to explain, "if you think we'd expect you to come into this knowing how to start IV's and do spinal taps, and knowing what to do in a cardiac arrest, you're out of your minds." Meeting blank stares, I went on: "All of us were interns once and we know how completely hopeless you are at this point. We know that all four years of medical school gives you is a basic foundation on which to build. Every medical school graduate knows a bunch of facts but very little practical information. You know all the complex physiologic mechanisms that are necessary for the digestion of food by the intestine, but you've never actually taken care of a patient with a malabsorption syndrome; you know how the glomeruli of the kidneys filter impurities out of the blood, but you've never had to manage fluids and electrolytes in a patient whose kidneys have failed. That's what you're going to do in this internship: learn how to put all these principles into practice. And while you're learning this stuff, we're not going to let you do anything that might even come close to hurting the patients. The only thing we'll ask of you over the next few days is that you somehow figure out how to get yourselves onto the wards without getting too lost. Anything more than that is extra credit. Now I'm sure that made you all feel a lot better, right?"

It obviously didn't, and they all fell silent again. "So what else are you worried about?" I finally asked. "Or is that it?"

"Well, okay, so maybe you don't expect us to be able to make decisions on our own, but there are a lot of other things we're going to be responsible for," Andy responded after a bit more silence. I liked Andy right away. "I mean, starting Saturday, parents are going to be trusting us with their sick kids. They're going to expect us to take care of them and make them well. I'm worried I'm going to wind up betraying that trust."

The other two considered this. "That's certainly a frightening thought for society," Mark added. "People trusting me with anything."

I added "Anxiety about responsibility" to my list.

"I don't know about you guys, but I'm worried about my home life," Amy Horowitz said next. "I can understand worrying about doing a good job, but I've got a two-month-old baby at home. If I'm on call every third night and I'm exhausted the next night, that means I'm only going to have one night out of every three to spend with her and just about no days."

I knew Amy from her days as a medical student, and of all the interns in the incoming group, she was the one about whom I was most concerned. "Who's going to be taking care of the baby while you're at work?" I asked.

"We have a baby-sitter during the day and my husband

will be home every night," she replied. "I've been on vacation since I delivered, and I've spent a lot of time with her. It's really going to be hard."

The others considered this and were silent for a few moments. "Yeah, outside life, that's a problem for me, too," Andy finally added. "I've seen what happens to interns. They don't have time for anything. They turn into boring, out-of-shape slobs, and I don't want that to happen to me."

And I added "Anxieties about home life: No time for families, hobbies, or exercise" to my list.

We spent about an hour talking. Even though it started out slowly, our discussion rapidly picked up steam. The interns had some kind of chemistry that made them work well together. By the time Mike Miller, the department's director of education, finally called us to lunch, our list of anxieties covered nearly two pages.

Things had gone so well during the second half of that hour that at the end of our session I told Amy, Andy, and Mark about a project I'd been thinking about for some time. "I've thought about trying to write a book about internship," I told them, "and I'd like you guys to help me with it. All you'd have to do is keep a diary and meet with me for dinner every once in a while. After some discussion, the interns agreed that they'd like to give it a try. Since one of their anxieties was that they'd lose touch with their nonphysician friends who had no understanding of what it was like to work a hundred hours a week and who therefore could not possibly sympathize with this lifestyle, they thought that a book about what an intern's life was like might be helpful to future interns. They decided that tape-recording their experiences would be the best method of keeping a diary.

And so what follows is the edited, collected diaries that

resulted from the suggestion I made on the lawn of Peter Anderson's house. As witnessed on that evening in the emergency room in May, Andy, Mark, and Amy came a long way during that short year: They were transformed from medical students into competent doctors. And each survived their year of internship with their lives largely intact.

. But what happened between that day in June 1985 in Westchester and the day one year later when it was all over is an amazing story, a story that, for people who have not lived through an internship, might seem more fictional than real. But all that follows is true.

To live by medicine is to live horribly.
                                   —*Linnaeus*

The Intern Blues

Andy

Sunday, June 30, 1985*

I suppose I should have started this diary forty-eight hours ago, before I'd actually started my internship, but I only got this tape recorder today. So now I've actually had a day and part of a night of call. I think I'll start by talking about what I think about being an intern.

The fact that this was going to be starting didn't really hit me until I began packing up my stuff last week. The last couple of months have been the best time of my life. I finished all my medical school requirements back in March, and after that, Karen and I took off for Asia. We traveled around for two months, having a great time and then got back to Boston near the end of May. I loved that time we spent traveling. It gave me time to think about the future. But then a couple of

---

*Like everything else, the interns' months are different from other people's months. In order to allow continuity of care, the interns switch to new rotations a few days before the calendar month begins.

weeks ago, I started getting ready to move, and that's when I really began to think about being an intern. I've had some pretty weird feelings about all this, and I guess I should try to put them into words.

Leaving Boston has been very traumatic for me. Except for the time I spent at college, I've lived around Boston since I was ten years old. It's a great city; I really got to know the place like the back of my hand. I met Karen a couple years ago; she's a fourth-year medical student now, and we lived together this past year. To leave all that, the city, my friends, my girlfriend, our apartment, my family, it's been a really a difficult, dramatic thing. It's going to be a big adjustment. The only thing that's making it a little easier is that Karen came out with me; she'll be around for another week or so, just until I've had a chance to get myself settled. But after that, she'll be gone, too. I don't know what I'm going to do after that.

I've felt kind of lost since getting to the Bronx. I moved in last Sunday night, and Monday and Tuesday, the first days of orientation, were really stressful. I came home on Tuesday after the lectures on management of emergencies and cardiac arrests and I just fell into Karen's arms and cried for a while. I've never done that before; it kind of scared me. I felt so wound up about these new responsibilities that were looming larger and more threatening every second. I felt terrible. I just thought, "What am I doing here? I can't do this, I'm not good enough to know how to rush in there and save people's lives when they're dying all over the place, when they are bleeding and not breathing."

Something about those lectures scared the shit out of me. It wasn't that I hadn't heard the stuff a million times before; it was the way the lectures were saying it. "Well, *you'll* want to do this and *you'll* have to know this, because *you'll* be the intern." They weren't talking *at* us, like they did in medical school, they were talking *to* us. That was scary. I really didn't sleep well those first couple of nights, mostly because there was so much to think about.

I guess one of the things I'm worried about is how much this internship means. When I started medical school, I had great expectations about how much I'd know and how skilled I'd be when I graduated. I thought I was going to be a doctor with a capital "D." Now that I've finished medical school and I've been through all the disillusionment about the capabilities of twentieth-century technological medicine, I feel like I don't even deserve to be called doctor with a small "d." Medical school turned out to be a very negative experience, a real grind for the first two years, sitting in lecture halls day after day, week after week, being bored to death by people who don't seem to care about anything except what's happened in their research lab over the past ten years, and a mixture of wasted time, humiliation, and feeling intimidated the last two years. As a third- and fourth-year medical student, you get to realize how unimportant you are, how things go on whether you're there or not; you're only there to get yelled at by the attendings for doing stupid things, or to get abused by the house staff, who treat you like a slave. You don't really learn how to be a doctor in medical school. So I'm coming into this internship hoping and praying it's everything medical school wasn't. I'm hoping again that when I finish this part of my training, I'll be that doctor with a capital "D," but this time there's more pressure on me: This has got to be it. When I leave here, I've got to be a doctor.

And I'm entering with expectations that this'll be an exciting and interesting time in my life, with memories I'll always cherish. I know it's not going to be easy, and it's not going to be a lot of fun. I'm going to feel lousy a lot. But I hope when I'm all done, I'll be able to look back at these years and be able to say that the time was better spent as an intern than in almost any other way I could have spent it. I've invested four years in this already; if I invest another three years and wind up realizing I hate being a doctor, well, that's seven years of my life completely wasted. I'm twenty-seven, I'm still a young guy, but seven years of wasted time, that's pretty sad.

And finally, after all these emotions and worries, it started. I was on call on Saturday in the Mount Scopus emergency room. Once the day began, it really wasn't so bad. There was just a resident and me. My first case was a little eight-month-old with a really bad case of cervical adenitis [**swelling of the lymph nodes in the neck due to infection**], and I thought, wow, here I am, a real doctor, with real pathology. I wound up having to admit the kid for IV antibiotics.

I did pretty good during the day, I was really enjoying it. I wasn't scared, I wasn't freaked out and I don't think I made any really horrible mistakes. I went at my own pace, which was about half the speed of the second-year residents but I felt good about it. I did an LP [**a lumbar puncture, commonly called a spinal tap, a test in which a needle is inserted through the back and into the spinal canal, so that a sample of spinal fluid can be obtained for analysis**] on an 18-month on whom we had a suspicion of meningitis, and it went perfectly. I got the spinal fluid and I started an IV without any problem at all. I did a CSF cell count [**counting the number of white and red blood cells in the spinal fluid specimen in order to diagnose meningitis**], and I learned a bunch of good bench lab stuff that I never knew how to do before. Hell, it was a good day and we even got a chance to eat dinner. I got out of there at 12:30 A.M., which isn't bad. I have to say my first night on call was a positive experience, which gave me a good feeling about coming to this program in the first place.

Sunday, July 7, 1985

Karen went back to Boston today. Even though she'll be back in three weeks to visit for a weekend, I know things are not going to be the same as they were for at least this whole year. I took her to the airport, she went through the gate, and I stood there waving and she waved back until all I could see was her arm. Then that disappeared and she was finally gone.

I got back from the airport and putzed around the apartment for a while, feeling aimless. I dropped off Ellen O'Hara's [**one of the other interns**] car keys—she had loaned me her car for the weekend—and Ellen and I talked for a while. She was a little spacey; she'd been on in the NICU [**neonatal intensive-care unit**] last night and didn't get any sleep. Then I went out shopping.

I was in the vegetable store and I had this really funny feeling, like I couldn't think clearly for a minute. I couldn't figure out what was wrong at first, but then I realized that I was shopping for myself. I started feeling really bad because I'd be the only person in the apartment eating this stuff. When I got back from the store, I called Karen right away. She was home already. I told her how much I missed her and how lonely I felt. She told me she felt the same way. We talked for a while and when we got off the phone, I felt real down, real down, and I didn't know what to do. I just walked around the apartment, feeling very empty. I felt like I wanted to cry, so I sat down at my desk but the tears wouldn't come.

I had to talk to someone; the only person I knew was Ellen, so I called her and told her I'd like to come and talk for a while and she said sure. I went up to her apartment, she opened the door and asked what was wrong, and I told her I was feeling real low. We sat down on her couch and I started crying. I kind of fell onto her shoulder and cried for ten, fifteen minutes, really crying, soaking her blouse. She held me and I held on to her. I felt a lot better after that. We talked about getting together for dinner, and so I went back down to the store to get more food.

That was amazing! That kind of thing, crying on a total stranger's shoulder, is not something I've ever done before. I was feeling bad, really bad, and she was the only person I even knew here. All I can say is, I'm glad there are people like Ellen in this program.

But all is not lost. When I went out for food the second time, I found a store that sold Häagen-Dazs ice cream! Häagen-Dazs in the Bronx! Amazing! [**Häagen-Dazs ice cream has always been manufactured in the South Bronx.**] There's hope for this place after all!

Work is good. I've finished my first week as an intern, and it's shown me that I actually like being a doctor. I enjoy the people I'm working with, I like the kids . . . I'm rediscovering some of the things that made me go into pediatrics in the first place. This week, working in the ER [**emergency room**] and the clinics, I saw more kids than I had seen during the entire six-week rotation I spent in pediatrics in medical school. I love the kids, but I can see that the adolescents can drive you nuts!

There are a lot of things I don't understand about adolescents. Do you examine them with their parents in the room? Do you throw the parents out, and if so, when do you throw them out? And there are all these hidden agendas going on between the parents and the kids. The other day, I saw a fifteen-year-old girl with a vaginal discharge. Her mother insisted on staying in the room the whole time. I felt pretty uncomfortable asking the girl whether she was sexually active or not with her mother standing right there next to her, but the woman just wouldn't leave. So I wound up doing a pelvic exam and getting all the cultures and stuff without even knowing what I was looking for. I guess when I get some of these issues sorted out, I'll feel better about them, but as of now, give me those toddlers and little kids anytime!

I'm starting to feel more confident and more willing and able to see patients without supervision. [**In the beginning of the year, interns working in the emergency room are supposed to check with the attending on duty before discharging any patient.**] It was so busy the other day, I didn't have time to check everything out. We were about four hours backed up most of the time and I was just running from one thing to another without any time even to think, let alone consult an attending. Occasionally I asked for advice just to check myself, and the attending who was on call in the ER always agreed with what I wanted to do. That felt good; it was a real boost to my ego.

There is one thing about work that's starting to bother me, though. When I was in medical school, one of the things I liked best about pediatrics was dealing with the parents. Over the past few days, though, I've found myself getting really annoyed with some of the parents who bring their kids to the emergency room. For instance, this mother brought her two-year-old in the other night. When I asked her why she was there, she said, "My kid hasn't gone to the bathroom for two days." All she had was minor constipation, for God's sake! And the woman hadn't even tried anything. Here it is, twelve-thirty in the morning, and there's an eight-year-old boy in the other room who just got his eye blown out by a firecracker. And there's a five-year-old sickler with a fever in painful crisis [a child with sickle-cell anemia, a common inherited disorder mainly occurring in blacks and Hispanics, in which the red blood cells collapse when the blood oxygen is low; the collapsed, or sickled red blood cells clog up the smaller blood vessels, leading to obstruction and further lack of oxygen, which results in pain in the hands and feet; in a patient with sickle-cell disease, fever can be a sign of serious, possibly overwhelming infection], it's time for me to get home, and this woman brings her child in for minor constipation. I wanted to strangle her, just put my hands around her neck and strangle her! And this kind of thing isn't unusual. It happens all the time, every night, that's why there's always a three- or four-hour wait to be seen. I tried to be nice to her, but I can't help getting really pissed.

Occasionally there'll be parents who seem really weird. The other day, this woman came in with her two kids who had colds. She was like a street person, she was carrying around all her possessions in shopping bags and she was dirty and her hair was all matted. The kids seemed perfectly okay though. There wasn't much I could do; just examine them, tell her to give them Tylenol, and send them on their way. But it bothers me; there has to be something wrong there. I've tried to figure out a way to get kids like that away from the parents, to protect them, but it seems to be impossible to do unless the parent actually harms the kid. You can't call the BCW [the Bureau of Child Welfare, the state-run agency that investigates physical and sexual abuse of children] just because a mother looks and acts a little funny. Even when the parent actually does harm the kid, like when they beat the kid with a strap because he or she misbehaves or acts up, it's sometimes difficult to do anything to prevent it from happening again.

Tuesday, July 9, 1985, 3:00 A.M.

Must internship really be like this? Must it really have hours like this? Today was just one of those long, zooish days. I had clinic this morning, had about three seconds for lunch, went to the emergency room where there was already a big pile of charts in the triage box, and that's how it stayed until a little while ago. There wasn't even a minute to get some dinner; I was starving, but there just wasn't any time to stop. We kept working and working and the triage box of charts of patients waiting to be seen just kept getting bigger and bigger.

I've noticed I'm not nervous anymore. I did another spinal tap today, on a little one-year-old with a fever who had had a seizure. It went fine, the kid didn't have meningitis, it must have just been a febrile seizure [**a convulsion that occurs with fever and having no adverse long-term effects**]. I admitted a kid with anemia, the second kid I've admitted since starting. I used to sweat like a pig when I had to do procedures and stuff; I don't seem to be sweating much anymore. It happened very suddenly. So far, internship has been a period of exponential learning. I just hope all I'm picking up sticks.

I still can't believe I'm getting paid for this. But I'll tell you, I don't think it'll be long before I start thinking I'd damned well better be getting paid for this. I think that happens when you start to respect your skills. I'm not there yet; but I'm getting there, I think.

But I do get really pissed off about working in the West Bronx emergency room [**West Bronx, also referred to as WBH, is a municipal hospital adjacent to Mount Scopus**]. I was drawing blood today from a four-year-old and I had to stick him three times because he kept pulling his arm away and pulling out the needle. The reason he kept pulling his arm away was because the nurse wasn't holding him tightly enough. When I told her, she said, "I don't care, I don't give a damn!" Oh, really! She just didn't give a shit about the kid! Here's a woman who must really love her job.

I forgot to talk about something I can't believe I haven't mentioned yet. Something really significant happened tonight, something horrible, and I guess I blocked it out of my mind for a while. As the triage box was filling higher and higher with charts and we were getting farther and farther behind, we were called by a frantic clerk to come over to the critical care room. He said there was a pediatric cardiac arrest going on.

So we tore over there to see what was happening. I got there first. I found the place jammed with doctors and nurses working on what looked like a pretty big adolescent. They were pumping on his chest, they had him hooked up to the cardiac monitor, they were sticking him for blood and starting big IVs in his groin. I had no idea what to do. The resident showed up a few seconds after I got there and we stood around for a couple of minutes until they just told us that we could leave unless we wanted to run the code. "No," we said (laugh), "it looks like you guys are doing just fine." But no one had taken a history yet, or even talked to the mother, so the resident told me to go out there and get the story. I found the woman; she was perched outside the critical care room looking scared to death. I took her over to the social work office and started talking to her.

Briefly she told me the kid was a fifteen-year-old asthmatic who'd been in the middle of a bad asthma attack when it sounded like he had become obstructed [**the main breathing tube, the trachea or one of the mainstem bronchi, the tubes leading from the trachea to the lung, became blocked**]. He stopped breathing and they loaded him into a car and sped off to the hospital. They were headed for Jonas Bronck but on the way the kid was snatched up by a passing EMS team and brought to West Bronx. He had been pulseless, breathless, and unresponsive for God knows how long. When he got in the ambulance, he had vomited and aspirated [**leaked stomach contents into his lungs**] and gone into arrest.

So he was kind of dead when they brought him in, but I don't think I really believed it. His first pH was 6.9 [**indicating severe buildup of acid in the blood, a condition resulting from lack of oxygen delivery to the tissue and not consistent with life for longer than a few minutes**], which isn't great. His heart was beating only about eight times a minute, but he was a kid, and kids just don't die like this. Not the ones I'd known anyway.

When I was getting the history, the mother asked me, "How is he, Doctor?" and I was about to say . . . I don't know exactly what I was about to say, but then the clerk opened the door and took the mother away because he had to register the kid or something administrative like that, and I left, after telling her I'd come back to talk to her again when she was done.

Next thing I knew, that clerk came back to me, not as

excited this time, and he said, "The kid died; he's dead." I couldn't believe it. I knew he hadn't been doing well and that they were doing everything they could for him, but dead? I just couldn't believe it. I had to walk in and see him myself.

In the critical care room, the crowd was gone; there were just a couple of nurses, removing all the lines and stuff, cleaning him up, getting ready to bag him, and there he was with his glazed corneas—yeah, he looked dead, all right. The medical resident came in and we talked about it for a minute. No one had said anything to the family yet. I told him I'd gotten the history from the mother. "Well, I guess you're the only one who's established rapport . . ." he said. Rapport? I spoke with the woman for five lousy minutes; that's not exactly what I'd call establishing rapport.

But I was elected. Other than me, nobody had even laid eyes on the woman. The medical resident said he'd come along with me. On the way back to the social work office, I stopped myself and thought, What the hell am I going to say to this woman? I knew she was totally unprepared for this. When I had talked with her earlier, I got the impression she thought everything was going to be okay. I knew things weren't okay. I had seen him getting his chest pumped, being a full code. I should have said, "Your son is in critical condition. There's a chance he won't make it." I wish I had said it when I'd had the chance, but then that damned clerk had come in and had taken her out to register her. I should have booted him out, told him I was talking and that it was important, but I didn't think to do that, so I didn't get to prepare her in any way. Ah, maybe she didn't want to know, maybe she would have been worse off had I tipped her off beforehand. Who knows?

Anyway, there I was, sitting in front of her in the social

work office, and the medical resident was standing behind me and there she was, looking at me, not having a clue what was going on. All I could think to say was, "I'm sorry, but I have to tell you, your son is dead."

She looked at me, her eyes bugged out, and she became completely hysterical. And the woman who was there with her also became completely hysterical. They began screaming in Spanish and wailing and throwing themselves around. I didn't understand a fucking word they were saying, I didn't know what was going on, they were making a tremendous ruckus and I just . . . I just didn't know what to do. It was a terrible moment. I felt completely powerless. I couldn't think of anything to make her feel better. It was probably the most horrible moment in her life.

As we were walking to that room and I knew I was going to be the one to tell her about her son, I remembered hearing about situations like this, when you have to tell a mother that her child has died, and you don't even know her; you're just on call and it's not your patient and you just kind of get signed out to take care of the dying person. I expected it to happen sooner or later; I'm just kind of surprised it happened so soon in my internship, in virtually the first week.

Wednesday, July 10, 1985

I spent the afternoon in the West Bronx ER, where I had a great case. We had this kid I saw a couple days ago, the one-year-old who came in with a febrile seizure. I tapped him and found he didn't have meningitis, but today the blood cultures I'd sent came back positive, with gram negative rods [**meaning that there was a bacterial infection in the blood with a bacteria called *E. coli*, a potentially serious infection**]. We called the kid back in and he still had fever on the antibiotics I had prescribed, so we admitted him for treatment of sepsis **[infection of the blood]**.

Then later in the day, a little five-month-old came in with a high fever. I did another spinal tap and did the cell count and this time, yes, lots of white cells; this kid did have meningitis and was admitted. That's routine pediatrics, very routine, I know, but for me it was a very exciting thing. I was able to do everything from start to finish, and that was all stuff I learned here, stuff I didn't know how to do in med school, and that's very exciting. I'm now able to do some things that doctors are supposed to do.

I told the mother we'd have to admit the baby and take more blood and she said "No way." She was frightened to death. I knew there was no way that I, with my vast two-week experience as an intern, was going to be able to convince her to let us do what we needed to do, so I called the attending and he came over and sat and explained the whole thing to her, telling her how important it was that we start an IV and begin antibiotics as soon as possible so that the baby would have the best chance possible of surviving and she listened carefully but it was clear she was so frightened she couldn't think straight. She finally said, "I don't know what to do anymore. Call my husband and ask him." We called the father and he said, "Do whatever you have to," and he came in. It was very sad: Here's this beautiful, normal baby with this terrible infection and the real possibility that he'll wind up retarded, and I was excited because I had been able to do the workup from start to finish. It's like I'm less interested in the patients and more interested in what I can and cannot do.

I've been here a week and a half and I've done more spinal taps than I'd done in all of med school. You see a lot in this place, it's a great program, but I can see how I could get burned out. It's a real danger in a place like this, with call relentlessly every third night and the ancillary staff somewhat less than helpful. I can see I'm going to have to watch out; down the road when it's the winter and my vacation is a month or two away, I can see how I could possibly come to hate this, how what seems like fun and is exciting now could turn into a real drag later.

Monday, July 15, 1985

Time seems both to drag and to race onward. It seems like forever since I last saw Karen; it's only been a week and a half. I really miss her.

Last night I was on call in the West Bronx ER again, and from the word go, it was crazy, packed from the start until about two o'clock. I didn't have even a second to catch my breath. This is getting to be a very disturbing routine.

I spent Friday and Saturday down in Manhattan with some of my friends from college, Gary and Maura. They live in the city; it was nice to get out of here and see some people outside of medicine. I tried to explain to them about some of the stuff I've been doing and seeing. They say they understand, but I get the feeling they only believe about half the things I tell them.

Today I acted as the supervising physician for IV sedation in a kid who was getting a radioulnar fracture [**a fracture of the two bones of the forearm**] reduced by the orthopods [**internese for orthopedic surgeon**]. Although the attending popped his head in a few times to make sure everything was all right, I basically just did it on my own. Even though nothing really happened, it's still kind of a nerve-racking experience.

Thursday, July 18, 1985, 2:00 A.M.

Just finished another call in the West Bronx ER. The past few days have been mixed. Today was pretty good, but the two days before pretty much sucked. I had a couple of aggravating days in clinic [**all pediatric house officers are assigned to a "well-child clinic" in one of the hospitals; interns and residents have office hours once or twice a week in clinic during which they usually see six or seven of their own patients**], where I just felt overwhelmed and disorganized; it was driving me crazy. The problems were pretty boring, but I'm picking up lots of new patients, slowly but surely drumming up my clinic. I have the feeling it's going to be a booming clinic pretty soon. It seems like every walk-in [**a patient who comes to the emergency room**] needs a regular doctor. They ask me if I'd be their doctor, I say sure and give them the clinic's number. I have a feeling this is going to be a mixed blessing in the long run, but anyway . . .

I was really tired most of today. I just don't seem to have any pep. It's this every-third-night-on-call business, the inevitability of it, it's just dreadful. Even though this is the easiest rotation I'll have and I get to go home every night (even though it might be at three or four in the morning), these hours just get very tiring. Is it possible that I'm really starting to get tired this early into the year? I'm worrying about everything; I've even started to have trouble sleeping on the nights I'm not on call. I didn't sleep well last night—I woke up three times before my alarm went off.

Well, it's time to go to sleep, my favorite pastime.

Saturday, July 20, 1985, 3:30 A.M.

Today was my worst day of internship so far, because of two incidents I had with orthopedics. First, there was a kid with a dislocated elbow. I was doing the IV sedation and the prick ortho resident didn't like the amount of sedative I was giving, he wanted the patient to get more and at one point he actually put his hand on the syringe full of morphine I was holding and started to squeeze. I had to shake his hand off and tell him, "No, you're not supposed to do that." The rest of the procedure was punctuated by him cursing at me for not wanting to give enough sedation. The jerk!

Later there was another kid who needed sedation, so this same resident and I decided together to give him a DPT [a **cocktail of three sedatives: Demerol, Phenergan, and Thorazine, given through an intramuscular injection]**, but the nurses wouldn't do it. They have this rule that DPT is not allowed to be used. So this started a big stink and things were getting more and more hairy. The pediatric resident who was on got pissed off at the nurses and they got pissed off at us, and the ortho resident's yelling, "Hey come on, guys, hurry up!" Finally we decided to give IM [**intramuscular]** morphine but I wrote the order on the wrong part of the chart and the nurses didn't see it and they didn't give the medication and before you know it, the ortho resident was back, pissed as hell because we were taking so much time, and he started yelling at me for being so incompetent and then I started yelling back at him and I could feel the blood rising in my face. I've never felt that angry at anybody before. It was making me crazy that I had no way to get back at him, so I just kept yelling at the fucking guy, telling him he was a jerkoff and a dickface. It was a very uncool thing to do.

Right after this, I grabbed a chart and went into an examining room but I was still so angry, I couldn't concentrate. So I told the senior I needed two minutes to cool off, and I went down to the vending area to get a Coke. I put my money in the machine, and what came out? A nice, warm Pepsi! No ice! No refrigeration! Oh, God, how I love West Bronx!

I went to a corner, sat down, and tried to cool out for a while. Then I went back to the ER, got some ice, and drank my fucking Pepsi. I apologized to the nurse I yelled at; I even apologized to the ortho resident, even though I think I'd still like to break his arm.

A few of us 'terns got together the other night and went out. We had dinner at an Indian restaurant in Manhattan, then went to get some ice cream and roamed around for a while in the rain. It was pretty good, but we were all so damned tired. Everyone was either postcall or precall. Shit! It's just amazing how often call comes around. It's like you feel you just got off and it's your night again.

Wednesday, July 24, 1985, 12:20 A.M.

The month's almost over and I'm looking at the end with mixed feelings. On the one hand, this ER stuff is starting to get pretty old. It gets repetitive and frustrating after a while. But on the other hand, I can't say I'm looking forward with any great excitement to being in the NICU next month. I basically like to work decent hours; I don't like staying up all night, which is what you have to do in the NICU. Thank God there're caffeine and other stimulants.

Actually, I've never taken other stimulants. I never liked the idea of speed. Of course, I've never had to go for more than twenty hours without sleep. Being exhausted and having a little twelve-hundred-gram baby crumping [**trying to die; deteriorating**] in front of you, that kind of gets you worried.

Sunday, July 28, 1985

I had a really wonderful weekend. I was on call Friday night and it was really quiet. I got home at twelve-thirty and Karen was here. She had just gotten in. We stayed up and talked until I was too tired to stay awake and then we went to sleep. It was restful, relaxing, and wonderful. Then today we were down in Manhattan and I saw Karen off to her bus to the airport and I started feeling very sad again. I've felt kind of sad and kind of nervous and lonely all day.

I took the train back home. That's the worst part for me: coming home to an empty apartment, knowing no one's going to come home after me, that I'm not waiting for anybody, I'm home and that's it. I called a couple of people; I called my friend Anne from medical school. She's an intern in Boston. We had a nice talk. Then I called my mom, who had just gotten home from England, and we talked for a long time. Then I got ahold of my intern friends Ellen and Ron, and we went out for some dinner. That was nice; I needed the companionship, I needed to be with people I felt close to.

And tomorrow morning I start my new rotation in the neonatal intensive-care unit, and I'm on call the first night. I've heard all sorts of horror stories about being an intern in the NICU.

It's only been a month since I started, and I can already see a big change in myself. I don't think anyone outside of medicine really understands what this whole thing is about. I've had trouble explaining my life this month to people, and I'm sure next month is going to be even more impossible to explain. I'm starting to think that it probably isn't even worth the effort for me to try. Most people in the nonmedical public, they have their own ideas about what doctors should be

like, and I don't think they want to have those ideas shat-
tered. They don't want to know about the long hours and
the lack of sleep and everything else. They have these myths
that we're all like Dr. Kildare or Marcus Welby. I hate to dis-
appoint them by telling them the truth.

Amy

Tuesday, July 2, 1985

The hardest part of this year is definitely going to be leaving Sarah. There's no question about it. I'm not sure yet what being an intern is going to be like, but I am sure of one thing: There's nothing they can make me do that could possibly be any harder than saying good-bye to the baby was for me yesterday morning.

Some people might think it a little strange to have a child two months before you start an internship. Well, there are a lot of things that went into our decision. First and foremost, Larry and I have wanted a baby ever since we got married. We both love children; that's really the main reason I decided to go into pediatrics in the first place.

Another reason has to do with my family. My mother died when I was in college. I still haven't gotten over it. Ever since, I've wanted to have a baby, a girl, and name her for my mother. That's been very important to me. That's why our baby is named Sarah.

I guess the third reason has to do with my miscarriage. I was pregnant when I was a third-year student. That one wasn't planned; I just got pregnant. Larry and I were both happy about it. I went to my obstetrician's office when I was about ten weeks and he heard the fetal heart. Everything seemed to be going fine, but then two days later, I started to have some pains and Larry had to take me over to the Jonas Bronck emergency room, where the miscarriage was diagnosed. They did a D and C and sent me home. If I hadn't lost that pregnancy, I'd have had the baby in the beginning of my fourth year of med school, which would have been perfect: I would have been able to take some time off then; things aren't too hectic in the fourth year. And I'd have had a one-year-old at this point, and leaving a one-year-old all day with a sitter isn't as bad as leaving a two-month-old. But having had that miscarriage, I started to wonder whether there was something wrong with me. I thought I'd never be able to have a baby. I guess I became obsessed with it.

Well, all of these are reasons for having a baby, but they don't explain why I decided to have one two months before I started internship. I guess the reason I didn't want to wait until after I finished my internship and residency was that you can't tell what might happen; there are people who wait and something happens to them medically and they find out that they can't have a baby. I didn't want to take a chance. I had thought for a while that maybe I'd take this year off and spend my time just being a mother and wait until next year to do my internship. A lot of people advised me against that. I was told it would be hard to get back into medicine after I'd been away from it for a whole year. And Larry encouraged me to go ahead with my internship; he told me he could manage the baby when I was on call. So here I am.

But I didn't think it was going to be this hard to be away from her. From the day Sarah was born until orientation started last week, I spent every minute with her. I took May off as vacation time. In the beginning of May, we put an ad in the paper for a full-time baby-sitter. We chose the woman we finally hired because she seemed really nice and she had great references. Her name is Marie; she's a Jamaican woman who's about forty. She has full-grown children of her own. She started two weeks ago, while I was still around. She's going to come every morning, Monday to Friday, at seven-thirty and stay until Larry or I get home at night. Larry and I will be alone with Sarah on weekends. Marie seems to like the baby a lot, but then again what's not to like? There are a couple of little problems, though: She spends all day carrying Sarah around, she feeds her every two hours because she says she's afraid the baby will cry and get colic. I guess I'll eventually have to talk with her; I'll have to be more assertive. I know Marie'll do a great job and everything'll turn out okay. It's just that . . . I'm worried she might turn Sarah into a wimp!

So far, being an intern isn't any worse than being a medical student. I'm in the OPD [**Outpatient Department**] at Jonas Bronck this month. I was on call last night for the first time and I got out of the ER at about a quarter to three. There was a lot of trauma, plenty of lacerations and head wounds, but since I don't know how to do anything yet, I wound up seeing the more basic medical problems. For some reason, most of the kids I saw were four-month-olds with fevers. There's something going around, I guess. I felt bad for Evan [**the senior resident who had been on call that night**]. He was the only person who knew how to suture, so he wound up spending the entire night sewing lacerations.

Since the attending went home at ten o'clock, the other interns and I had to keep interrupting him every five minutes to discuss patients with him. I felt bad doing it, but I wasn't about to send anybody home without clearing it with a senior first!

When I got home, something weird happened. I went to bed and I must have fallen into a deep sleep because Sarah started crying at around four and I thought I was still in the ER getting ready to see another screaming kid. Larry told me he heard me say, "Please God, let me go home."

Monday, July 8, 1985

I've been on call three times now. Last Friday was the worst so far. Everything had been going pretty well until about eight o'clock, when a thirteen-year-old girl who had been raped came in. I wound up seeing her.

She was a young thirteen; she looked more like eleven or ten. She was really broken up, but I got her to tell me what had happened. She had been alone in her family's apartment when a knock came on the door. She looked through the peephole and saw her fifteen-year-old brother's friend. He told her he had left a book in the apartment, and she let him in. They went into the brother's bedroom and started looking around. Suddenly the girl felt something around her neck. The boy had pulled out an electrical cord and he kept pulling it tighter until she got down on the bed and took off her clothes. He then proceeded to rape her.

He was there a total of about an hour. A little while later the girl's family came home and found her hysterical. They immediately brought her to the ER.

The attending and I went over what had to be done. I did a complete exam and got all the samples that would be needed as evidence when the case went to court. There's something called a rape kit that has to be used, with directions that have to be followed exactly or the whole thing can be thrown out of court. I made sure I did everything right. I was working like a robot all through it, trying not to think about anything except getting the job done. After I finished, I handed the rape kit over to the cop. The social worker came in to talk to the girl and her parents, who were crazy at that point. The father wanted to go out and kill the kid slowly, really make him suffer. The mother just cried. The girl didn't speak much, she was in shock. The cop called a little while later to say the boy had been caught. The parents took the girl home at about midnight. As soon as they left the ER, I just fell apart. I spent the next hour crying. We were still busy, so when I pulled myself back together I had to start seeing patients again. We didn't get out until nearly five in the morning. It was a terrible night. Terrible.

Yesterday was busy, too, but it wasn't nearly as bad as Friday. Yesterday's specialty was trauma. We had all kinds of trauma, kids falling out of windows onto their heads, firecrackers blowing off fingers, and the basic foot laceration from Orchard Beach [**a beach on Long Island Sound**]. Since I still haven't learned how to suture, I spent most of the day seeing kids with head trauma. Most of them were okay; I just examined them, found nothing wrong, and sent them home with a head trauma sheet [**the emergency room provides instruction sheets in English and Spanish covering most of the common pediatric problems**]. At about 3:00 A.M. we had finally cleared out the triage box and I picked up one of the last charts, a six-year-old who had hit his head on a coffee table. The nurse who had seen him when the mother first registered at about eleven thought he looked all right and put him at the bottom of the pile. [**The nurses triage each patient according to his or her symptoms: Patients who require real emergency care are "up-triaged" and their charts are placed near the top of the pile; patients who are judged to be stable are triaged to the bottom of the pile. On exceptionally busy days, the wait to be seen by a physician may be as long as six hours.**]

The mother told me the boy was running around the apartment and had fallen and hit the back of his head on the coffee table. He hadn't blacked out but had become very sleepy. I examined him: He *was* sleepy, but then again, so was I. There were no focal findings. [**A focal neurologic exam, one in which there is weakness, paralysis, or abnormal reflexes on one side of the body, indicates a neurologic deficit. A negative neurologic exam following head trauma is a fairly good indication that the brain hasn't been harmed.**] The senior resident told me it was okay to send the kid home, but when I went to give the mother a head trauma sheet, the kid suddenly couldn't remember anything that had happened over the past few hours. So the senior told me to check the kid out with the neuro fellow [**the pediatric neurology department trains a group of fellows, individuals who have completed two years of residency and have gone on to do another three years in neurology**]. It took him twenty minutes to answer his page, and when he did and I told him the story, he told me he had to check with his attending [**the senior doctor on call for neurology that night**] before deciding what to do.

It took him another forty minutes to get in touch with the attending! There I was in the ER in the middle of the night with nothing to do but wait. I could have been home sleeping! But I couldn't sign this out to the night float. I had to stay.

When he finally called back, he said the kid needed a CT **[CT scan: a computerized X ray of the brain]** and that he had to be admitted. I had to bring him up to the fourth floor and help the technician get him settled on the CT table. Finally, the intern from the ward showed up and I got out of there. I didn't get home until four-thirty. If I hadn't picked up that last chart, I would have gotten about two hours' more sleep.

Thursday was July 4 and I had the day off. It was great: just me and Larry and Sarah. We went swimming in the pool at our apartment complex; Sarah seems to love the water. It was like getting reacquainted.

I had another run-in with Marie on Friday. I'm still having trouble with her. Even though we had a talk, she still carries Sarah around all the time and feeds her every two hours, as soon as she opens her mouth and lets out a peep. We got into a fight last Monday. I came home for lunch after clinic ended at about eleven-thirty. When I came into the apartment, Marie was holding the baby. I tried to take Sarah, but Marie wouldn't let go of her. I just about had to pry them apart. I said a few things I probably shouldn't have said. I don't know what's going to happen with her.

I think having a baby at home is making me into a more efficient intern. I find myself trying to get my work done as quickly as possible and running home. Sarah gives me a lot of motivation to work fast.

Friday, July 12, 1985

I feel bad talking about this, but I think I should anyway. Maybe it'll make me feel better to get it off my chest; I don't know. I feel terrible and it isn't my fault, it just isn't my fault.

I did what I was supposed to do, that's all I know, and somehow I got into trouble.

Last week, this adolescent girl came in complaining of rectal bleeding. I recognized her from my subinternship [**a two-month rotation in the fourth year of a medical student in which the student works as an intern, taking night call and admitting and following patients**]. She was on the ward for some psych problem, I don't remember exactly what. I examined her and did a rectal exam; I didn't find anything wrong, and the stuff I smeared from the glove onto the guaiac card was negative [**the guaiac test is for hidden blood in the stool**]. So I thought she was a crock [**a patient who has nothing wrong and is faking symptoms; short for a "crock of shit"; sometimes referred to as a "turkey"**]. I presented her case to Tom Kelly, who was the attending, and I told him . . . I told him . . . I'm sorry, I have to stop for a minute and pull myself together. Crying is not the answer.

Sorry. Anyway, I told Tom I thought the girl was lying. He told me that very well might be the case, but that I should draw some blood for a CBC [**complete blood count: a blood test that measures anemia**] and a sed rate [**erythrocyte sedimentation rate, a test for inflammation or infection, in this case used to rule out inflammatory bowel disease, which is a cause of rectal bleeding**] anyway, to rule out any real problem. I agreed with him and I went back and got the blood. I sent the CBC to the lab and spun a hematocrit [**a direct test for anemia**] myself. It was fine—37 or something like that. I forgot to get the sed rate, though. I just . . . it slipped my mind. The girl . . . she was making me crazy while I was taking the blood and I just forgot about it. I did it, I don't know why they don't believe me. . . . I'm sorry, please forgive my outburst.

A few hours later, Tom got it in his head that I hadn't taken the blood. I don't know why. He didn't come to me directly and ask. He checked with the clerk and she told him she didn't remember sending anything off to the lab. Then he called the lab and they denied ever getting a CBC on the patient. Then he called the girl at home and asked her if anyone had taken blood from her and she denied it. And then he came and confronted me with all this and accused me of having lied about the whole thing. I don't know . . . I did it, that's all I can say. I don't know why the lab didn't get it. The girl's crazy, so I can understand her saying I didn't take it. I don't know. . . .

The worst part of this is that now everyone knows. I didn't do anything wrong, I swear it, but all the ER attendings are talking about it. Some of the residents know. Nobody's going to trust me, I know it. I've been crying for the past three days. I'm on tomorrow and I can't let this affect me. I have to go on and just ignore this. But how can I do that?

Wednesday, July 17, 1985

Things are a little better today. On Monday, Jon Golden **[one of the pediatric chief residents; the chief residents are responsible for the house officers; they directly manage the patients, make up the on-call schedules, and look after house staff morale]** came to talk to me. He told me he had heard about what had happened in the emergency room and he assured me he didn't believe it and that I shouldn't let it bother me or interfere with my work. It was nice to talk to him, it made me feel much better, but it's sad to think it had reached the higher-ups that fast. Even though we've got so many interns and residents, this is a small department in terms of gossip. You have to be really careful what you tell people.

Then yesterday, Tom Kelly talked to me in the ER. We cleared the air. He apologized for the story getting around. I understand it in a way; the ER attendings at Jonas Bronck are all close friends and they have to talk to each other about something! Anyway, it looks like this storm has passed. Now if I could only straighten things out with my baby-sitter.

I'm still having trouble with Marie. I don't know what to do about her; she's very good but she's got her own ideas about things and there isn't anything I can do to change her. It comes down to two options: Either I try to live with the way things are, or I let her go and try to find someone else. I don't think it would be very easy to find a replacement for her right now. I barely have enough time to eat; I don't think I can afford to go through another round of interviewing. So, I guess I'll have to keep her. It could be worse.

I wonder what the other interns are doing with their free time. I mean, I spend all my time away from the hospital with Sarah. I never talk to anybody outside of work.

Friday, July 26, 1985

Yesterday was a terrible day. I was on call. I seem to be on call every night! But yesterday was worse than usual because it was Sarah's birthday. She was three months old and I wasn't there to see her. In the last couple of weeks, she changed into a real person. She doesn't just lay in her crib anymore: She can lift up her head and look around and she smiles when she sees me. She's got a beautiful smile. It's great coming home to see her. I just wish I could spend more time with her!

We were really busy all day and I couldn't seem to do anything right. In the afternoon, I had this four-year-old who came in with a high fever and chest pain. She was coughing and congested and I was sure she had pneumonia. I listened, and sure enough, I heard rales [**a crackling sound in the lung fields, usually associated with pneumonia**] in the area of the right lower lobe. I sent her for an X ray but it turned out normal, no evidence of pneumonia. I went to the attending, Harvey Abelson, for help. He examined her and said he didn't hear any rales but he found that she had right CVA tenderness [**tenderness over the costovertebral angle, the area of the back that overlies the kidneys**] and told me to check the urine. Sure enough, the urine was loaded with bugs [**bacteria**] and white cells. I gram-stained it [**a test to identify the type of bacteria causing the infection**] and found *E. coli* [**the most common bacterial cause of urinary tract infections**]. She had pyelonephritis [**an infection of the kidney**], not pneumonia. She needed to be admitted, and I had almost sent her home! I'm just about finished with my first month of internship, and I think I know less now than I did when I started.

I didn't get home this morning until after four and I didn't get much sleep. I'm exhausted. But what can I do? This is the way internship is, and I've just got to survive it. At least I've got tomorrow off. Then I'm on again on Sunday for the last time this month. Larry's team has a softball game on Sunday, so he's going to take Sarah with him. Thank God he's so good with her! He takes her everywhere he goes when

I'm on. I don't know what I'd do if he weren't so understanding or helpful. If I do make it through this year, it's
because of Larry.

   I start at University Hospital on Monday. I don't know
what that's going to be like. I've heard bad things about it. I
don't care if it's hard or boring or whatever. All I care about
is getting out at a reasonable hour.

Mark

Friday, June 28, 1985

My internship officially starts tomorrow. I've waited for this day a long time, years, the way people wait for their arteries to clog up enough for them to have a heart attack. I'm starting on 6A, the ward at West Bronx. I talked to some of the old interns and they told me that 6A was a pretty horrible place to work, that the nursing stinks, the lab technicians were impossible, and, on a good day, the clerical staff mostly just ignored you. Sounds like my kind of place! I like a challenge like that. It's just what I need, a little more aggravation. And of course I'm on tomorrow night, the first day of the year. Just my luck; I'll probably also be on call every holiday and the last day of the year as well.

At least orientation's over. It was a lot of fun, if your idea of fun is hanging around a bunch of terrified lunatics who are just figuring out that they've made a terrible mistake in their career choice and have ruined their entire lives. It really wasn't that bad; at least the food was good.

It's funny, my coming here to Schweitzer. I worked with AIDS patients in New Jersey during med school and I became pretty convinced it was a disease I didn't want to experience personally. So I was looking for a program that didn't have a lot of AIDS patients, and Schweitzer didn't exactly top the list of places meeting that criterion. I went on some other interviews and got a look at some of the other programs and suddenly the Bronx didn't look all that bad to me. So whatever happens during this internship, I can't blame anybody but myself; it's all my fault.

Carole wasn't all that happy about my coming here. I guess I should mention Carole. We've been going out off and on since we were seniors in college. She's an accountant in Manhattan and she lives in New Jersey, so my being in the Bronx is probably the worst thing that could happen to our relationship. We'll probably never get to see each other over the next year. Not that it would be much better if I were working in Manhattan. It's pretty hard to keep up a reputation as Mr. Romance when you're working a hundred hours a week. Oh, well; being an intern is probably going to be like becoming a monk; except monks have a stronger union, I think.

Well, I'm going to try to get some sleep now. I'm sure I'll wind up walking around the apartment half the night. I've got all these butterflies in my stomach, and it feels like they've just organized a softball game.

Monday, July 1, 1985

What a great idea it is to start new interns on Saturday! What better way to greet someone who's not only completely new to the hospital and doesn't even know where the bathrooms are, but also who has never worked as an intern before, than to have him cover a ward filled with twenty-five sick patients, none of whom he's ever seen. I wonder who came up with that brilliant stroke of genius?

Needless to say, Saturday was a complete disaster. I started off the day just nervous, but by the time we finished work rounds at about ten, I was completely petrified. I mean, they had kids with meningitis who could die without batting an eyelash, kids with asthma who were on oxygen, and they were telling me to do things like "Get a blood gas on that kid" [**blood gas: an analysis of the acid, oxygen, and carbon dioxide levels in the blood; usually performed on patients in respiratory distress**]. "Check the X ray on that kid," "That other kid needs a new IV," etc., etc. I had to sit down for an hour after rounds, just to talk myself out of quitting right then!

I have to admit, the technical stuff is about the only thing I can do. So I started off by drawing all the bloods and starting the IVs that were needed. Then I was heading for the lab, wherever the hell that was, when I got called down to the emergency room to pick up a patient. And that was the first time I got lost. I couldn't believe it, I wound up wandering around in the basement of the hospital for twenty minutes, having no idea where I was. I found the morgue, I found the Engineering Department, but the ER seemed to be missing. I had this medical student with me, but he wasn't much help; this was the second week of his first rotation and he was more confused than I was, if such a thing was possible. We finally found a guy down there who spoke English and I asked him where the ER was. He laughed at me for a few minutes and then told me I was on the wrong floor, it was one flight up.

Anyway, my first admission was a six-year-old with asthma. My first admission. What a moment! I wanted to have the kid bronzed so I could hang him from my car's rearview mirror, but his parents wouldn't give their consent. I managed to find my way back to the ward with the kid, who didn't seem all that sick, and then I got yelled at by the head nurse: "You have an admission? Nobody told me about any admission! We're not ready for an admission!" So I made a mistake, but she stood there, blocking the door to the treatment room like she wasn't going to let us in. What'd she expect me to do, bring the kid back down to the ER and call and tell her I was coming up with him? The way things were going, I probably would have screwed up, gotten off on the wrong floor again, and wound up bringing the kid straight to the morgue (at least I knew where that was).

Well, I finally apologized, told her I wouldn't let it happen again, and she let us into the treatment room. There really wasn't much to do for the kid except take a history, do a physical, and write an order for aminophylline [**an asthma medication**]. Even I could do that! So it only took me about an hour to finish, and then I tried to get up to the lab but I got called back to the emergency room for another admission.

This went on all day. I got one admission after the other from about noon until after seven at night—six admissions in all. By that point I had a lot more lab tests to check and finally made it up to the thirteenth floor [**the laboratory floor at West Bronx**] at about eight. It took me an hour to check all those labs. Then I came down and had to show them to the senior resident and he told me what to do next. So it was about nine o'clock, I had gotten six new patients, I had done most of the scut that had been signed out to me, but I had missed three complete meals, I hadn't even had a chance to pee (it was about then that I felt the top of my bladder hitting my rib cage). And that's when all the IVs started falling out.

I don't know what it was, but all of a sudden three nurses came up to me at once and told me that an IV had come out on one of her patients. Three IVs at once! It seemed to be too much of a coincidence. I went to find the senior resident to ask him if it was possible that the nurses were pulling them out to torture me. He said it definitely was a possibility but there was nothing I could really do about it and, no matter what happened or what I was thinking, I'd better not get into a fight with any of the nurses or my life would be ruined. I told him about what had happened when I had brought up my first patient and he just sighed and shook his head.

I got two more admissions in the middle of the night and more IV's fell out and there was more scut to do and I didn't get a chance even to lie down but somehow I made it through and nobody died. So I guess, all in all, I'd have to say it was a successful night. My only problem is, I don't ever want to be on call again!

Well, after a night like that, at least today was pretty good. I met our team's senior resident, and he seems great. His name is Eric Keyes and he's got a weird sense of humor. Then I met our attending, Alan Morris, who's director of pediatrics at West Bronx. He's very serious and kind of stiff, but he also knows a lot and I've heard he's great.

I didn't get out tonight until after nine. I was trying to get my work done, but I kept getting lost around the hospital. I'm really hopeless. I'm going to have to get better organized.

Saturday, July 6, 1985

I am definitely on the chief residents' hit list. I've been on 6A for a week now, and everything seems to have settled down. I've finally figured out where everything is; I know where the admission forms are kept and where the lab slips are stored. I've found the ER three straight times without getting lost. I liked the people I was working with. I even made up with the nurses; I brought them a box of cookies on Tuesday, and amazingly, no IVs fell out when I was on call Tuesday night. Everything was going fine and then, yesterday, just when I was really beginning to feel comfortable, one of the chief residents came up to me and said they'd decided to transfer me to the Children's ward at Mount Scopus. I told her I didn't want to go, that I was having a really good time on 6A. She said she was sorry but they were a body short on Children's and there was an extra person on 6A that month (because we had a subintern) and there was nobody else who could be pulled. I argued a little more, but I could see there was no way I was going to change her mind. Finally I just gave up. So after figuring out 6A, I had to move over to a completely different ward in a completely different hospital, pick up a whole new group of patients, learn where the Mount Scopus labs are and where the forms are kept, and I have to meet a whole new group of nurses and probably go through another night of IVs falling out. Terrific!

Children's isn't so bad, though. First of all, Elizabeth is there this month [**Elizabeth Hunter, one of the other interns, went to medical school with Mark**], and it's nice to be together with her. And the other intern, Peter Carson, seems like a nice guy. And everybody tells me that Children's is much easier than 6A. I can see that that may be true: I was on last night and actually got three hours of sleep. Three hours of sleep! It was the first time I've even seen an on-call room since I got here. There are fewer admissions, and the patients aren't as sick. The chiefs have promised me that they'll still give me the month of Children's I'm scheduled for later in the year, so maybe this will work out well. I doubt it, though; in internship, nothing is supposed to work out well.

I spoke with Carole a little while ago and told her I missed her. I think she was surprised to hear me say that. I'm not what you'd call the most demonstrative person around, but I really do miss her. I miss everybody! All I've done since starting this internship is work or go home and fall asleep. I don't know if it's possible to survive a year like this.

I'm going to sleep now. Maybe I'll dream that it's next June and all this is finished.

Thursday, July 11, 1985

Yesterday was my third night on call on Children's. When I was on last Thursday and Sunday, I didn't have any trouble. I didn't get a single admission, and I slept three or four hours each night. I was thinking maybe I'd try to arrange to go to Children's for my vacation this year. So I wasn't at all prepared for what happened last night.

I got five admissions. There was one every hour or two. I'd just have time to finish working up one when the next one would show up. They weren't really very sick: an eight-year-old who came in to have a repair done on his cleft lip that had originally been repaired when he had been a baby; a kid who had developed an infection in his leg after he had been bitten by a dog and needed IV antibiotics; another kid who had periorbital cellulitis [an infection of the tissue surrounding the eye; dangerous because it can lead to infection of the eye itself and, occasionally, of the brain] and who needed

IV antibiotics; and an asthmatic. The sickest kid was also the most interesting; it was a four-year-old with a week's worth of facial swelling.

He had kind of an interesting story: His mother noticed the swelling around July 1 and brought him to their local doctor. The LMD [**local M.D.**] sounds like he graduated from a medical correspondence school. He brilliantly decided the kid was allergic to trees and started him on Benadryl [**an antihistamine used to reverse effects of allergic reactions**]. The mother was back at the guy's office in two days: Not only didn't the Benadryl work, it also seemed to make the swelling worse. Now not only was the kid's face swollen, but also his hands and feet were puffy. The LMD told the mother that sometimes it takes a while for the allergy to get better, especially since there are so many trees around, and that she should give the medicine more time to work.

Well, she gave the medicine as long as it took for the kid's belly and scrotum to get swollen and then she brought him to our ER, where one of the residents made the diagnosis in less than a minute. The kid, of course, had nephrotic syndrome. [**This is a disorder in which the urine contains large amounts of protein and, as a result, the body becomes protein depleted, leading to swelling of the body. The face, especially the area around the eyes, is typically the first area affected in children. The disorder usually is self-limited; it is treated with steroids and usually resolves in weeks.**] For God's sake, even I could've made the diagnosis! So we admitted him, I called a renal consult, and we started him on 'roids [**internese for steroids**].

One of Elizabeth's patients almost got kidnapped two days ago. This three-year-old had been brought to the hospital by her aunt last week. The kid had a couple of episodes of blacking out while the aunt was baby-sitting for him. A whole workup was done and everything was negative except the tox screen [**toxicology screen, a blood and urine test looking for toxic substances in the system**], which was positive for alcohol. So social service started an investigation. The mother and her boyfriend showed up on Tuesday at noon, and after visiting with the kid for a while, asked to speak to the doctor. Elizabeth showed up and they asked her how the baby was doing. Elizabeth said that he was fine, and before she could say another word, the mother said, "If he's fine, I want to take him home."

So Elizabeth told her he couldn't leave yet, that tests were still pending and that, for the sake of the child's health, he'd have to stay at least one more night. Then the mother started yelling that if her baby was fine, the only thing that could happen to him in the hospital is that he could get sick, which was actually a good point, and she picked the kid up and started moving toward the elevator.

At that point, we all moved in. Someone called security stat [**immediately**], and within a minute a phalanx of Mount Scopus's finest emerged from the elevator bank and we had a standoff. The mother held on to the kid tight and shouted, "I don't want my baby in this fucking hospital!" at the top of her lungs, which went a long way to put most of the other parents on the floor at ease. Next she yelled, "I know what's best for my own fucking kid! If he's fine, I'm taking him home! Just try to stop me!"

Attendings, house staff, administrators, and more security guards started to show up. The mother and her boyfriend got madder and madder. The boyfriend finally said, "We're taking the kid out of here! If you don't like it, you might as well shoot us in the back, 'cause we're going!" The kid was screaming at the top of his lungs while this was going on.

The whole thing lasted about a half hour. It ended when an administrator, obviously someone who had majored in psychology and guerrilla warfare in administrator school, showed up and firmly told them that maybe they'd like to talk the whole thing over in the conference room. For some reason, the mother agreed and she, the baby, and the boyfriend headed off with him. I think our suspicions about the parents were correct. The BCW **[Bureau of Child Welfare, the state agency charged with investigating child abuse]** probably will be interested in doing an investigation.

I've nodded off to sleep three times while recording this. I think it's time to stop.

Wednesday, July 17, 1985

I'm a little more coherent tonight, I think. Nothing much is happening. Elizabeth's patient whose parents tried to kidnap him got sent home by Social Service last Friday. In their infinite wisdom, they cleared the family in two days. I've got a bad feeling about this family. I hope I'm wrong.

My patient with nephrotic syndrome is doing much better. Most of the swelling is gone, and he doesn't look so much like Buddha anymore. Those steroids are amazing! We're going to send him home in a few days; renal will follow him as an out-patient. They say his prognosis is excellent. The mother asked me if all this means he's not allergic to trees. I told her I thought it probably would be a good idea not to go back to that LMD anymore.

Last night was pretty easy. I got four hours of sleep, and that's been pretty much the pattern on Children's. I guess I did kind of luck out when they switched me from 6A. Those guys have been getting killed. As far as I know, none of them have gotten any sleep on any night they've been on call.

Wednesday, July 24, 1985

I'm about ready to die. I thought I was bad that night earlier in the month when I was up all night, but this is ten times worse. I haven't gotten any sleep for the past two nights, and I'm pretty worried about my grandmother.

I haven't mentioned my grandmother yet. She's my mother's mother. She's over eighty and she lives in New Rochelle by herself. I try to get over to her apartment for dinner at least once a week, usually on Tuesdays, if I'm not on call or too tired. I went last night and I found out she was really sick.

She's got a bad cellulitis on her leg. She cut herself with a knife about a week ago. When I showed up yesterday, she was febrile and looked terrible; she could barely get out of bed. She showed me the cut; it was all red and swollen with lots of pus. Her temperature was 102.5, and I told her she had to go to the hospital for IV antibiotics. She said I was crazy. She's a little on the stoic side. I argued with her for about an hour and finally convinced her to let me take her to the Mt. Scopus ER to at least get a third opinion. I got her seen without any wait. A medical intern looked at her and said, "You've got to come into the hospital for IV antibiotics." She started to tell him he was crazy, but I guess maybe she really wasn't feeling so well because she finally said, "All right."

She's on one of the medical floors. They put in an IV and started her on megadoses of pen and naf [**penicillin and nafcillin, two antibiotics**]. They didn't get her settled until after two in the morning. I stayed with her until six and then went home to change my clothes and take a shower. I might as well just move my stuff over to the hospital. As it is, at this point I'm only just occasionally visiting my apartment.

Anyway, I don't know how I got through work today. I've got seven patients, and I don't remember what happened to any of them. I was like in outer space for most of the day. My mother showed up this afternoon to stay with my grandmother, and I came home. I'm going to sleep now. I remember sleep; I think it's something that feels really good.

Friday, July 26, 1985

The past few days have been nothing but a blur. I was on last night and I managed to get some important sleep. My grandmother's much better; they're probably going to send her home over the weekend. And my time on Children's is coming to an end. Of course, I'm on the last day of the month. You can almost set your calendar by my on-call schedule. And then on Monday, I start on Infants'. I have the feeling the shit's about to hit the proverbial fan. Infants' is a bitch!

The only good thing about all this is that I know I'm not going to be on the first night. The chiefs may have decided they don't like me for some reason, but they're not crazy. They couldn't make me work two nights in a row. But actually, since I'm on Tuesday, I get a weekend off next week. Weekends off, I remember those; that's when you get to visit your apartment for two whole days.

Bob

You might wonder how these three interns wound up coming to our little corner of the world. It is not fate or destiny that brought them here, but rather the bizarre intern mating ritual known as "the Match."

All of medical school—in fact all of life—is nothing but preparation for the Match. It's the first of many horrendous and inhuman experiences to which house officers are exposed. In other professions, a person who wants a particular job submits an application and a résumé; the person goes on interviews, trying to convince the employer that he or she is right for the job; if the job is offered, the person has the right to accept it and begin work, or to reject it. But this system, good enough for American business, apparently is too simple for medical residency training. After all, there's no torture involved.

The search for the perfect internship begins early in the summer before the medical student's fourth and final year of school. The student interested in pediatrics or internal medicine fills out as many as twenty applications for residency programs. He or she then spends a month interviewing at hospitals around the country, asking numerous questions of the house staff and attendings, trying to get a feel for the place. After narrowing the field down to a few top choices, the senior arranges to do "high profile" rotations at these hospitals. These rotations, often a subinternship in an ICU setting, give the student the opportunity to work himself or herself sick, taking call every third night, in hopes that somehow the director of the program will notice and think highly of him or her and possibly place the person near the top of the match list. But I'm getting ahead of myself.

Here in the Bronx, a committee of pediatric faculty members is attempting to select an outstanding group of interns from a pool of hundreds of applicants. For our entering group of thirty-five, more than 225 senior medical students were interviewed in the fall of 1984. This interviewed group was ranked from one to 225 on the basis of grades, letters of recommendation, the impression made during the applicant's interview, and performance during these elective rotations spent at one of our hospitals.

The fun of the Match actually begins in January. Each applicant sends off a list of programs to which he or she has applied, ranked from first choice to last, to the National Intern and Resident Matching Program (NIRMP) in Illinois. Simultaneously, the director of each program submits a list ranking all senior students who have applied for a position. All this information is fed into a computer and the machine grinds out the Match, coupling applicants and programs. One might think that this chapter of the matching procedure would end with a friendly letter mailed from NIRMP and received anonymously and privately in a mailbox some days later. But no; nothing in an intern's life is that simple!

The results of the computer's work are stored in a vault and released in the middle of March. The senior students from each school are assembled in a centralized location, one usually designed to maximize feelings of anxiety and hopelessness, and the envelopes are distributed one by one by the person, usually a dean of the medical school, charged with guarding the secrecy of the Match. A name is called, the student rises and slowly approaches the front of the room; the envelope is handed over, it's cautiously opened, and the student either sighs a sigh of great relief because his dream has actually come true, he's matched at his first choice program and as a result his future is assured, or he lapses into an immediate and frightening anxiety attack, often complete with hyperventilation, because he's gotten his third, or fourth, or, God forbid, fifth choice and is going to have to work at a hospital with a bad reputation or, worse yet, at a place that's considered "anti-academic" and no matter how hard he works in his internship, his residency, or his fellowship, he truly believes that he will never be able to become a true success.

Those anxiety attacks are fueled by a fact known to all subscribers of the Match. Unlike normal job offers, the Match assignments are binding. Unless there are major extenuating circumstances, there's no chance of changing once an assignment to a hospital has been made.

Why fourth-year medical students put up with this system has something to do with the whole mentality that supports internship. "It's the way it's always been done," "it's accepted," "there's nothing we can do about it," are the usual responses when the question of why it continues to be done this way is raised.

Well, that explains how the interns got into our program. I probably should next explain a little about the composition of our program.

The Schweitzer School of Medicine's pediatric training program is made up of two campuses. The one that's presently referred to as "the east campus" is composed of two hospitals: Jonas Bronck, a part of New York City's municipal hospital chain that provides primary care to the poor and not-so-poor of the northern reaches of the South Bronx; and University Hospital, a voluntary facility that mainly acts as a tertiary-care center for patients referred for consultation to the school's subspecialists by private physicians in the North Bronx and in lower Westchester County. University Hospital is located about a half mile south of Jonas Bronck.

"The west campus" is also made up of two hospitals: the Mount Scopus Medical Center, a huge voluntary hospital that, like University Hospital, serves as a base for subspecialists; and the West Bronx Hospital, sometimes referred to as WBH, another municipal facility that, like Jonas Bronck, provides all medical services for the indigent families of the western region of the borough. Mount Scopus and West Bronx are literally attached to each other. Although the Mount Scopus–WBH complex is immense, filling four square city blocks, the pediatric services in the two hospitals are adjacent to each other and conveniently connected by a bridge. The east and west campus hospitals are separated from each other by about five miles.

The program, with over a hundred house officers, 120 full-time faculty members, four chief residents, and over two hundred inpatient beds spread over the four hospitals, is one of the largest pediatric training programs in the country. Our interns rotate through three emergency rooms, six primary-care clinics, seven general pediatric wards, two pediatric intensive-care units, three neonatal intensive-care units, and three well-baby nurseries. If you're confused reading this, just think what it must be like for the interns who have to become familiar and comfortable with the nursing staffs, ancillary services, medical forms, and peculiar habits of the laboratory personnel in all these different hospitals before they can even think about taking care of patients.

So the question naturally must be asked, why would any-one electively want even to attempt to deal with all this? Internship is difficult enough, what with the long hours and the frequently depressing subject matter; what would possi-bly motivate someone to want to come to our program, where the difficulty seems to be compounded by the massive size and complexity of the place? Well, probably the main rea-son medical students want to train at the Schweitzer program is because of the amazing variety of experiences to which they will ultimately be exposed. Our residents see asthma and pneumonia, ear infections and lead poisoning, the mundane, "bread and butter" of pediatrics at West Bronx and Jonas Bronck, the municipal hospitals; but they also see the con-genital heart disease and the renal transplant patients, the craniofacial cases and the weird metabolic diseases, all of the rarer medical and surgical problem patients who wind up being referred to Mount Scopus and University Hospital, the voluntary hospitals in which the subspecialists lurk. So when a resident finishes three years in the Bronx, it can safely be said that he or she will have seen every kind of pediatric patient who exists. Our graduates know that nothing will ever surprise them; they'll have had experience with anything that might darken the threshold of their medical offices.

That's why Mark Greenberg came to the Bronx. He told me he wanted to get as much experience with as many types of patients as possible during his training. After meeting him for the first time at orientation, I got to know Mark a little better this month. He told me he had chosen pediatrics because it was the third-year clerkship he had enjoyed the most. He had liked it for the same reason most people are attracted to the specialty: He said it seemed to make more sense to watch sick children get well than it did to watch sick adults get sicker and die.

Mark told me his biggest problem with being an intern is that his brain is always tending toward entropy. Unless he tries very hard to keep his life controlled, he becomes exceedingly disorganized. Disorganized is not a great way to be during internship. All interns share a common short-term goal in life: to get out of the hospital as soon as possible. One must be very organized to accomplish that. If Mark continues to be disorganized, he might have to consider permanently moving into an on-call room.

There's something about Mark I noticed very early in the month. It's a funny thing: There are some people who look great after a night on call. No matter how many admissions come in and how little sleep they get, these people look unbelievably good the next day. Mark is definitely not one of these people. He had a couple of bad nights during July, and this was readily apparent in his appearance the next morning: His eyes were very droopy; his reddish-blond hair was uncombed and shot straight up in the air in all directions; and his clothes looked as if they'd been slept in, which obviously was not the case because Mark always claimed that he hadn't gotten any sleep at all.

Amy Horowitz didn't really decide to come to Schweitzer; she decided to stay here. She had been a medical student in the Bronx and had stayed on because she liked the program and felt comfortable with the people. She's always lived in the New York metropolitan area. Born in Morristown, New Jersey, she was her parents' only child. Her father owns an office supply business.

I've known Amy for a little over a year. In March, when she was in the ninth month of pregnancy, she told me that she'd thought a lot about being an intern and having a young baby but was somewhat concerned that she wouldn't have time to be both a good intern and a good mother. But she's convinced she can do it. It's because of this conflict that Amy's the one intern in the entire incoming group about whom I'm truly worried.

Early in the month, a crisis developed involving Amy. While working in the emergency room at Jonas Bronck, she was told by one of the attendings to get some blood tests on a patient. Amy swears that she drew the blood and sent it off to the lab. The attending, in checking on the situation a little later, could find no evidence that the lab had received the specimen or even that the blood had been drawn. He confronted Amy and, when she affirmed that she had done what was requested, he accused her of lying.

Whether this is true or not, lying about lab results is about the worst sin a house officer can commit. The implications are far-reaching. First, although our department is immense, word of mouth travels like wildfire, and within three days of this incident, rumors about Amy had already reached every member of the outpatient faculty. Second, and more importantly, whether she was guilty or not, Amy has lost a great deal of credibility. Interns have to be trusted. Although life-and-death decisions are always made by more senior physicians, such as attendings or chief residents, interns must be expected to function fairly independently with only occasional supervision when it comes to performing the more mundane, everyday types of activities, such as drawing blood, checking lab results, ordering tests, or making appointments for their patients. Amy's ability to function independently has been called into question. Whether she drew that blood or not, Amy probably will have an attending or senior resident perched over her shoulder at all times to make sure she does what she's supposed to do, at least for the immediate future. Amy is smart and a reasonably good worker, and within a month or so she'll probably make everyone forget that this happened. But if she screws up just one time, she's going to get nailed. And that could be it for her for the rest of this year.

I'm pretty sure Andy Baron doesn't want to be in the Bronx. I think he was one of those people who had a major anxiety attack when he opened his Match envelope last March and found out he was coming here. I don't think he objected because of our program. It's just that he never thought he would actually have to leave Boston.

Except for college at Princeton, Andy's spent his whole life around the Boston area. He returned to that city after college, attended medical school at Tufts University, and vowed that he'd never leave again. He told me he ranked Boston Children's Hospital first on his list, and he'd been led to believe that getting in there wouldn't be a problem. So you might say he was more than a little surprised when he found out he hadn't matched there.

I think leaving Boston will have a major impact on Andy. Back home he had a very structured and broad-based support network. His family and friends are there, and, most importantly, so is Karen.

Karen Knight is the woman Andy's lived with for the past year. Karen is a fourth-year medical student at Tufts; she's going to have to spend a good portion of the year there. Andy has told me repeatedly that their relationship is strong, that it had lasted through a lot of adversity in the past, and that he feels it will easily be able to weather this year of separation. It sounds almost as if he were willing it to be that way.

And what was waiting for Andy here in New York? Almost nothing; there are a few friends who attended college with him, but nobody close who would understand or be there when things start getting rough. Internship is hard enough when you have a lot of love and support to help you through; it's nearly impossible when you have to go it alone.

Andy

Thursday, August 1, 1985, 12:40 A.M.

I just got off the phone with Karen, the only nice thing in the entire day. Three days of the NICU done, three and a half more weeks to go. What can I say? It's another planet.

Saturday, August 3, 1985, 7:00 A.M.

I just woke up. I'm thinking about going back to sleep again, but I've got to get to work. Internship is turning out to be so much harder than I thought it would. The NICU is amazing; it's only about twenty-five yards from one end to the other, and there are four little rooms off the central nursing station. In each of the rooms, which are about ten feet by ten feet, they have five or six tiny babies arranged with all this massive equipment around them. It's claustrophobic and frightening because each of the kids is so sick. Being in the NICU so far has been a total shock.

I was on call the first day (Monday) and I actually got a

couple of hours of sleep. I was on call again on Thursday and it was a horrible, horrible night. We didn't get any sleep at all. And there were these three kids who kept trying to crump [**deteriorating; trying to die**] on us. We seemed to be doing a good job of stopping them, but then at about five in the morning, little baby Cortes decided to really crump. Cortes was one of the "ageless" preemies who live in the ICU. She was born fourteen weeks prematurely, weighing about a pound and a half, and she'd lived for four months right on the edge between life and death. We called a CAC [**West Bronx's and Mount Scopus's term for cardiac arrest; literally, "clear all corridors"; also called a "code"**]. I pumped on her little chest for about half an hour while everybody tried to put in IVs and get access. We called for epinephrine [**a drug that stimulates the heart to beat**], and we called for more epinephrine and we called for bicarbonate [**a drug that reverses the buildup of acid that occurs any time blood stops circulating**], and we tried to give bicarb intraosseously [**through a needle directly into a bone, usually in the lower leg; intraosseous meds are given only in dire emergencies, when an intravenous line cannot be established**] and we got a blood gas and the pH was 6.6 [**indicating that there's so much acidity in the blood that life is not possible**], and then the heart rate kept slowing down and we gave intracardiac bicarb [**through a needle passed through the chest directly into the heart; used as a last-ditch attempt**]. And the heart rate came up again. Unbelievable! It looked like she was going to make it, but her color still was really bad. We bagged her [**blew oxygen through an ambubag through an endotracheal tube and directly into the lungs**] and we pumped her heart, but then

she went into V-tach [**ventricular tachycardia, a preterminal heart rhythm**] and we gave her some lidocaine [**an anti-
arrhythmic drug, used to reverse an abnormal heart
rhythm**], and then the surgeons came and did a cut-down [**a
surgical procedure in which a vein is found and a catheter
is placed into it, ensuring direct intravenous access**], and
we pumped some albumin into the femoral artery. We got
another blood gas; it was still 6.6 and the kid had deteriorated into an agonal rhythm [**a heart rhythm signifying
impending death**].

So we stopped the resuscitation. We had been working on
her for about an hour, I guess. There was nothing more to do.
I left the room and went back to try to finish up the evening
scut before the morning shift came on. The baby died. And I
felt really, really shocked. I felt stunned, like somebody had
hit me over the head with a two-by-four. I had gotten so close
to that little baby. She was so sick and so tiny. She was the first
patient I ever did CPR [**cardiopulmonary resuscitation**] on.
It's a strange thing doing CPR on a baby that small. It's kind
of an intimate act. You've got your hands all the way around
the chest and you're trying to pump her life back into her.
You're trying to prevent her life from ebbing out of her. It
doesn't matter that the kid's got snot running out of her nose
onto your hands, it doesn't matter that she looks like shit, you
just want her to live so badly! It was terrible when she died.

Laura Kenyon, our attending, came in at about eight. She
took a look at me and asked if I was okay. I told her I was
fine, and she took me into the on-call room and kicked everybody else out. "Are you really okay?" she asked. At first I told
her yeah, but then I said I was really upset and I started crying. I was crying for that little baby whose life we couldn't
save. I told her how much I liked that little baby even though

I hardly knew her. I told her how I thought we were going to bring her back to life and keep her from dying. I told her I'd seen other people die when I was in medical school, but this was completely different. It's different when it's a baby. She told me it was okay to cry, it was okay to feel bad because it meant you really care about people, about your patients. She said that eventually you're able not to feel so bad, you can internalize it, but that you always feel something, because each death reminds us of all the others that preceded it.

She was really good. She let me get that baby's death out of my system. She told me I could go take a shower and have some breakfast. That was nice of her, but I didn't do it because I knew if I left the NICU, I'd get horribly behind in my work, and I knew that once that happened, I'd never get out of there.

After she talked with me, Laura had to go deal with the parents. She told the mother what had happened, and the woman started wailing. I turned around and saw Laura walk out of the unit. She had this expression on her face; I could tell she was really upset. She put her hand over her mouth; she was fighting off tears. For a second or two she looked really different, she almost looked like a little girl. And then she began to regain her composure and her face returned to normal. I listened to the mother's wailing for a while, but then I had to get back to work.

A little while later, I had to go back into the room where the baby died to draw blood from another patient, and there was Laura with the parents looking at the poor little dead baby, all swaddled and wrapped up. All day, I felt really down. Any time I'd think about it, I felt bad . . . really bad.

During the day, I was completely drained. The night had been such an emotionally exhausting experience for me, I was completely wiped out. It was so bad, any time I sat down, I'd start to fall asleep. Laura gave a really good lecture on physiology that I wanted to hear, but I just kept falling asleep. It was embarrassing; at one point, in front of everybody, she said to me, "You can go to sleep if you want, Andy." I wanted to pay attention, but I just couldn't.

Laura's the most amazing attending I've met here. She's tough, but I think she really cares. I think she loves her work and she wants everything to work well, so she's willing to put in the effort to make everything work on all levels. It's really exceptional, having someone around like that. I'm lucky to have her as my attending.

I got out of there around seven. I was too tired to do anything. I went out and got some food and ate dinner. By seven-thirty, I was ready to go to sleep. Karen called at about ten. We spoke for over an hour. I kept telling her how much I missed her. We didn't want to get off the phone; we kept thinking of something else to talk about. It's really hard being away from her this long. It's another four weeks until we get to spend any real time together again.

So anyway, the NICU is a very strange place. It's very exciting, physiology in medicine brought to its highest application. But when you think about it, it's also a very sad place because there's life and death involved; you take these little babies, most of whom would have been dead ten years ago, and there they are, just sort of cruising along. I think the best workers for a place like the NICU would be robots, or people who can blot out all their emotions and just do the work that has to be done.

The technical work you do in the NICU is pretty straightforward; once you've had some experience, you get very good at it. But the technical stuff is really the easiest part of the job. It's the decision-making that's the hard part.

Almost every day in there, we're called on to decide whether
to keep a baby alive or to let him or her die. I don't have any
of the tools necessary to make those kinds of decisions. I
don't have any experience with preemies, I don't know which
babies might have a reasonable chance of surviving and which
babies don't. All I can do is what somebody else tells me.

A lot of these babies don't even look human. They're really
fetuses. Take poor baby Cortes, for instance; she weighed
about a pound and a half at birth. I don't know, it doesn't
seem to me like we're doing anyone any favors by working so
hard to keep a baby like that going. We're just delaying the
time when the parents'll have to mourn their baby's death.

Saturday, August 3, 1985, 8:00 P.M.

I thought I'd make a little list here, not necessarily in
order of importance:

What's Right with My Life

*1. I'm in an excellent training program and basically enjoying
my work, despite the fact that I complain a lot.*

*2. When I'm at work, where I spend most of my waking hours,
I'm with people who, for the most part, I like, some of whom I'm
becoming friendly with, people like Ellen O'Hara and Ron
Furman.*

*3. In my nonwaking hours, I'm in an apartment that I basi-
cally like. It's not great, but it's sufficient, and I tend to sleep
pretty well because I'm not overly anxious, even though I have
lots of reason to be.*

*4. When I'm not working, I have some old friends around
whom I get to see.*

5. New York City is a great place to live with tons to do, and

I'm taking a lot of advantage of being here. I went down to Manhattan today, my only day off for the next two weeks. Oh, well.

6. I'm not depressed, something about which I worried when I came out here.

What's Wrong with My Life

1. I'm not with Karen, and I miss her a lot.

2. Even though I've made a few friends, I don't have any really good friends here. I miss having good friends around whom I can call and talk to about the things that are troubling me.

3. I'm not wild about this neighborhood. As time goes by I find more and more that I do like, but basically it's a kind of boring neighborhood that tends to roll up its sidewalks at about eight o'clock.

4. I miss my family—my parents, my brother, his girlfriend, they're all back in Boston. I used to see all of them very often; they were a source of great support, of great enjoyment.

5. I miss Boston. I really like it. It's much more hassle-free than New York, a more sane and easy place to live, and far less crazy and bizarre.

6. Sometimes I wonder if I'm in a program that has just too goddamned much scut and is too goddamned big. Sometimes I wonder if the great downfall of this program is the fact that we rotate through too many fucking hospitals and we have to spend so much time and energy on just learning the mechanics of survival on all the different wards that there's almost no time and energy left for stuff like relaxing, socializing, reading, sleeping, and just thinking constructively and thoughtfully about the patients.

So, those are my lists. Now that I think about it, they are basically arranged in order of importance.

Tuesday, August 6, 1985

Things are going all right, I guess. I got rid of a couple of patients. I have only three right now, and they're pretty stable. And I got a decent night's sleep last night. I really needed it; I basically collapsed at nine-thirty after I got home totally wiped out from another all-nighter without any sleep. So right now, things are looking up.

But Sunday night was one of the worst possible nights I could imagine. I was on with Larry, the senior resident, and we were both working our butts off. I spent most of the afternoon and evening doing shitloads of scut. At about one in the morning, I finished most of my work and went up to the well-baby nursery [**the well-baby nursery, maternity ward, and labor and delivery suites are on the seventh floor of WBH**] to try to finish all the physsies [**physical exams; all well newborns must be examined within twelve hours of birth**]. There were a lot of new babies, and I was plowing through them all. At about 3:00 A.M. I realized that the chart of the baby I had just examined was still over in labor and delivery, so I went over there to get it. Just as I got through the door, a nurse came running out of one of the labor rooms, yelling, "Get peds! Get peds stat!" She saw me and asked if I was from peds. I told her I was and she said, "There's a little preemie just delivered right in this room."

Great! This was just what I needed at three o'clock in the morning. I thought, Oh, my fucking God, what am I going to do? I had never been alone with a new preemie. So I turned to the unit secretary, yelled at her to call Larry stat, and then I ran into the labor room.

Lying at the foot of the labor bed was this little fetus. The midwife said, "I measured him. He's twelve inches long." [**A baby's gestational age in weeks is roughly equal to two times its length in inches. Therefore, this baby was probably at about twenty-four weeks of gestation.**] The baby was tiny but he was moving and I didn't know what the hell to do.

Last week I had gone to the delivery room with the neonatal fellow to see a micropreemie who had just been born. We knew about that baby in advance and we knew that it wasn't going to be viable, but the fellow had taken me to teach me about what's viable and what's not. That baby had no breast buds, his skin was gelatinous, his eyelids were sealed shut, and he was only ten inches long [**all signs of extreme prematurity**]. And the fellow said, "This baby is clearly not even twenty-four weeks; he's not viable. There's nothing to do for this baby." So we didn't do anything, and he died. And that had been my one experience with extreme prematurity.

Well, I checked all those things out in this kid. I measured him, and sure enough, he was twelve inches long. I looked at the eyes and they were sealed and there were no breast buds and the skin was gelatinous and I thought that this kid couldn't possibly be viable. Then I listened to his chest; he had a strong heartbeat, so I rethought the situation and figured maybe I was wrong. I didn't know what to think.

I decided to take the baby over to the warming table in the DR [**delivery room; all delivery rooms are outfitted with resuscitation equipment for preemies**] to see what I could do. Everything I knew was telling me that this baby could not possibly survive, but I just hadn't had enough experience and I was all alone. I ran into the delivery room with the baby and I laid him down on the warming table. I realized I didn't have any idea what to do next. I figured I'd try some oxygen: I grabbed the oxygen mask, turned the oxygen on and started to try to bag the baby, but the face mask was too big; it went over his whole head. I wasn't having any success.

Just then the baby kicked a couple of times so I listened to the heart again with my stethoscope. It was still beating pretty strongly. I decided that weighing the kid might help decide whether he was viable or not, so I asked the nurse to get a scale. And just then, as my panic was reaching its peak, Larry came walking in. Thank fucking God! I think I had been out of the labor room for maybe a minute by that point, but it had definitely been the worst minute of my life.

I told Larry everything that had happened. He took one look at the baby and said, "Forget it. This kid's not viable. Don't do anything." I was pretty relieved. I still felt bad because I didn't even have a clue about what I was supposed to do, but at least I realized I hadn't done anything that was harmful.

Then the nurse came back with this rickety old scale; it looked like something out of the nineteenth century. We put the baby on it and it read twelve hundred grams. No way that baby weighed twelve hundred grams! She said, "Well, this is the scale we use to weigh all the babies." Larry said, "Well, it's wrong."

We wrapped the baby up in a towel and brought him back into the labor room. Larry explained to the mother that the baby was too small to survive but since he still had a heart rate, we were going to have to take him down to the NICU. The midwife started throwing a shit fit. She said, "You can't take the baby downstairs! This baby belongs with his mother! You have no right to take the baby out of this room!" Larry told her that he wished he could leave the baby, but it was hospital policy that any infant with a heartbeat had to be brought to the NICU.

Then Larry and the midwife started fighting about where the baby should be kept while we waited for him to die. I stayed out of it; I agreed with the midwife, but I wasn't going to argue with the resident who had just rescued me. Finally Larry called the hospital administrator. She showed up, heard the story, and agreed with Larry. The midwife argued with her for a while but finally she backed down and we took the baby downstairs.

When we got down to the NICU, we reweighed him; he really weighed only 460 grams. We put him in an isolette **[also called an incubator—a Plexiglas box with a mattress and a heating element, used to house sick newborns]** to keep him warm. I checked his heart rate about every ten minutes. Finally, after an hour, the heart stopped and I declared him dead. Then I went upstairs and told the mother that the baby had died. We brought him back up and gave him to her to hold for a while. She was exceedingly sad.

Then I went downstairs and started doing more scut. At about seven o'clock all the new nurses came on, and they started yelling at me. They wanted to know why I hadn't filled out the death certificate and gotten permission for an autopsy. They were being really hostile. I was exhausted and I'd had a horrible night; all I wanted to do was be left alone. I didn't even know I was supposed to fill out the damned death certificate and get consent for the autopsy. Nobody told me I had to do those things.

Finally, one of the nurses came up to me, and she was

really nice. She knew I hadn't done any of this stuff before so she showed me exactly what had to be done. She gave me the death certificate and the autopsy form and the form for burial. She told me that I should go up and talk to the mother and tell her that if she wanted a private funeral, it'd cost $600, and if she didn't have the money, the city would bury the baby free.

So I went back upstairs and talked to the mom, told her how sorry I was. I didn't know what to say; I don't have a lot of experience with this. I asked her if she wanted us to do an autopsy and she said no. She was really broken up.

So I was up all night working pretty hard. Then today we rounded nonstop until one-thirty and then I had to go to my outpatient clinic. I signed out all my work. I saw five patients in clinic and that went pretty well. I got done by four o'clock or so and then I sat around and talked with my clinic preceptor, Ann Covington, for a while. I like Ann a lot. It's nice to have someone calming like her to talk to.

I went back to the NICU after clinic to finish my work. I got out at about eight and had to go to the bank and to the supermarket. The A&P closes at eight, so I missed it and now I can't go shopping again for another three days. I'm totally out of food. I have to bring stuff home from the deli down the street if I want to eat dinner. It's either that or going out every night. Fuck!

Karen called last night; I guess it was after I had gone to sleep. I don't even remember what we talked about. I don't remember a word I said. We talked for quite a while, I think. It's ridiculous. I hope she's home tonight so I can find out what's happening.

So here it is, eight twenty-five on my good night, my one night out of three that I'm not either on call or postcall and I have nothing to do and I have to go to sleep in an hour, so I can get a good night's rest before I'm on call tomorrow. I hate this! I think I'd really like the NICU if I weren't so tired, but I'm tired all the time. And you just don't get any normal human contact in your free time unless you're married or living with somebody. Even though I've made a couple of friends, they're all interns and they're either on call or tired. I really should be doing some reading about neonatology tonight, but screw it! I've got to get out of here!

Friday, August 9, 1985

I have a pretty nice white cloud right now [**white cloud: good luck on call; black cloud: bad luck on call**]. I still have only three patients. One's just a grower [**a preemie who has no medical problems except that he weighs less than two kilograms, the necessary weight for discharge from the NICU**], and the others are pretty easy also. Poor Dina, the junior resident, she's got five patients, three of whom are pretty sick, two of whom are *really* sick, both with NEC [**necrotizing enterocolitis, a serious disorder of the intestinal tract**]. I offered to take one of them but she didn't want to give them up, I guess. I'm on tomorrow, so I know I'll be picking up a sick kid who was born this afternoon, and I heard there's another preemie on the way, so I'll have at least two new ones to pick up. That'll fill out my service to five. Not exactly a piece of cake, but still pretty easy.

Boy, was I *dumb* on rounds today! Laura asked me a simple question about how much glucose I was giving one of my kids. Shit! I couldn't remember how to calculate it; everyone was standing there staring at me. I felt like an idiot. Later on I finally figured it out. Rounds are generally good, Laura's a great teacher, and except for when I'm making a dope of myself, I really enjoy it. Well, I've got to stop now, Ron's here, and we're going out to dinner.

Thursday, August 15, 1985

Being cooped up inside the NICU, you miss things and you don't even know it. I was riding down to Manhattan in the train this evening, you know, there's always something to look at, there's always guys coming through, telling you their life stories, begging for money, never fails. Walking around Manhattan on the way to the theater, I was just looking at all the people. They were all well-dressed, there were some very pretty women, something I almost never get to see in the Bronx. I realized that after only two weeks, I already missed the excitement that exists in Manhattan.

Today I got a call from Nelly Kahn, one of the social workers who works in the outpatient clinic. She told me she thinks I should report one of my clinic patients to the BCW. It was a mother who told me that she beats her kids with a strap when they act up. Ann Covington was right there, so I talked it over with her and she thought I should, too. So I had to call the mother and tell her I was reporting them. It really surprised me, she took it pretty well. Maybe it was like Nelly said, maybe letting us know she hit the kids was like her cry for help. I'll never figure these mothers out.

Then I called up the BCW, and they put me on hold for about twenty-five minutes! Twenty-five minutes, and I finally only got to speak to someone for five minutes. I was kind of surprised, the worker seemed really nice and friendly. I thought they'd be boring bureaucrats. All they wanted to know about was whether there were marks on the child. I told them there were and that we'd taken Polaroids of them.

They said that was enough, they were going to start an investigation.

Sunday, August 18, 1985, 2:00 A.M.

I've been in the NICU every day now for two weeks solid (having been on call last Saturday) and I finally have a whole day off. The sick thing is, I'm thinking I should go in today for about an hour because there's a workup I didn't quite finish. It wasn't really clear that I was supposed to be taking this one patient. It was one of those situations, I thought the resident was picking the patient up, then it turned out she wasn't . . . I don't know. So I may actually go in for an hour, just to finish that up, then I'll split before anybody catches me there and asks me to do something else.

It looked like Friday night was going to be really easy. All my notes were written early, and I was ready. It looked like I was going to get to bed by two in the morning, then things got complicated and then, around four o'clock, the deliveries started. Shit! Then it was just one delivery after the next. What do they do, wait until four in the morning to have all the deliveries? It's always like that! So the bottom line is I didn't get any sleep.

Truthfully, I've only actually gotten the chance to lie down in the bed in the on-call room once since the first night of the month, and that was only for about fifteen minutes or so. It was last week, at about five in the morning. The bed was unmade, the room was a mess, but it felt great! I fell asleep right away but I got woken up about fifteen minutes later: This weird guy who must have been high or something was in the room with me. He was opening and closing the door and doing all kinds of weird things. I lifted my head up and yelled, "Who the fuck is that?" and he ran away. I got up and locked the door, but my beeper went off and that was it for dreamland. Oh, well.

There were a couple of exciting things that happened the other day. There was a twenty-nine-weeker [**born eleven weeks prematurely**] who was an extramural delivery [**born outside the walls of the hospital**]. We got stat paged to the ER, so we went running down the stairs, and there's one of the pediatric residents holding this tiny, tiny baby. The guy looked uncertain about what to do. So we took the kid, who was doing fine at that moment, and we whipped him upstairs and wound up intubating him [**placing an endotracheal tube into his trachea so that direct ventilation of the lungs could be accomplished**] and so on and so forth. And it turns out the mom's a drug abuser. She claims not to use them intravenously, but who knows? So I might have gotten my first AIDS patient, although it's a little too early to tell, but who knows?

I went and talked to the mom later in the day. She doesn't want the baby at all. It's really sad. The father is nowhere to be found; when she was telling me this she got really teary-eyed.

So I didn't get out of there until six last night and I was just delirious. I'm not as good a doctor postcall as I am precall; I don't think anybody is. You just can't make as good decisions when you're that tired. I think postcall, I function at about 80 percent, which is not bad, but that extra 20 percent, that's got to be important sometimes. I think it's really stupid, I just think this whole unbelievable call system is stupid because it really makes you . . . you're just not as good! Don't misunderstand me, I'm not so much complaining that I'm unhappy about having to take the abuse of being up all night every third night, I don't like that, I don't like the way it makes me feel, but the thing that *really* bothers me is I don't think I can give as good care. If you're trying to give the best care in the world, you should be able to work out a system where doctors can function at their best. Anyway, I'm sure this won't be the last time I tirade about the evils of call.

Monday, August 19, 1985, 6:45 A.M.

I can talk only for a minute or two because I have to go back to work. I'm glad that I have only another nine days to go in the NICU and that I have only three more calls (one of which is tonight), because I don't think I'm wild about neonatology. I can't say it's been a horrible experience, but I wouldn't want to spend my life with tiny babies. There're much more interesting things in pediatrics than little tiny critters.

This morning I've been feeling kind of low; I've been missing Karen a lot. I talked to her yesterday morning, but she could talk for only a minute. I tried to call her last night, but she wasn't home. I really feel cut off. I fell asleep thinking about her and missing her and I woke up this morning feeling kind of low and lonely. I never want to do this again, be apart from her for so long, never, never. I never want to feel this homesick for Boston again either. It's eleven more days until Karen will be here, and she's coming for a month. It's going to be great, really great.

In the meantime, nine of those days I've got to go bust my butt. So that's what I've got to do. I'm on call tonight with Larry, the third-year resident in the NICU; I'm kind of glad about that. There's a definite difference between the second-year resident and the third year; the third years let

you do things on your own; the second years hog all the procedures. So with the second year, all you do is scut, but when
you're on with the third year at least you get to feel a little bit
like you're doing something. And Larry's a good guy; he's a
really fun guy, I'm sure we'll have a good time.

Well, I guess I gotta go. I'd rather go back to bed. But I
gotta go . . . I know I'll feel better about it when I get there;
it's always hard just getting there, though.

Friday, August 23, 1985, 7:20 P.M.

I'm in bed, and I'm going to go to sleep because I was on
call last night and I didn't get any sleep, and I'm really tired
because I worked my butt off. It's really ridiculous, this every
fourth is crazy . . . I mean every third. I suppose I should
have a lot to talk about . . . it's all so much of the same
shit . . . you know . . . creatinine, BUN, all that shit . . . it's
all gobbledygook. I'm going to sleep. . . .

Saturday, August 24, 1985, 8:00 P.M.

In another couple of days I'll have finished my second
rotation. Two down out of twelve and already, so early on, I
feel tired. I'm not worn out, I'm not whipped yet, but I feel
tired. I feel the effects of this every third night, it's already
wearing on me. And I already hate the system; I think it's a
stupid and foolish system that rules your life and hurts your
patients. And already I'm losing some sensitivity toward people—you know, as hurting, suffering human beings.

This morning when I woke up, I thought, What will I
do with myself today? I got this fantasy; I thought about
going back to my college, Princeton in New Jersey, and just
spending the day down there. I figured I'd look up a couple

of my professors and try to go see them. And then I got into the shower and the more I thought about it, the more appealing it got. I had it all planned: I'd go into the city and catch the train down to Princeton. I thought of all the beautiful green lawns and the tall trees reaching way up over the buildings, and about the flowers that would be in bloom, and the serenity and peacefulness of the place since there wouldn't be any students there yet. I became entranced with the whole idea, how quiet and pretty and pastoral it would be.

Then I got out of the shower and as I sat around thinking, I realized that I wasn't going to go to Princeton, it was all just a fantasy. My professors weren't going to be around, they were going to be out of town on vacation. And I couldn't just go and hang around there. I just wanted to escape from the difficult times I'm having right now into the past, when life was easier, when I didn't have to worry about all the diseases and falling asleep on attending rounds, the jaded attitude of some of the residents, the oppressed lives my patients and their mothers lead, the crummy neighborhood I live in and the fact that I'm far away from my loved ones. I think this'll blow over. When I finish this year and my residency and I'm just a practicing doctor when my hours are more regular and I'm more used to the responsibilities, it won't be so bad. I don't have any control over my life now, and that's very difficult.

I know I'm just saying this stuff over and over again, but it's just so difficult! I knew internship was going to be hard, everybody tells you that, you see the interns work their butts off, you know it's hard! But somehow you don't believe it. I was trying to tell my friend Maura about being an intern tonight, she really wanted to know. I told her a little bit, I told her a story, and then I just shook my head and all I could say to her was, "It's just really hard." I was thinking, Why am I trying to explain? I don't want to explain anymore, I don't want to tell anybody about this, it's just too crazed.

Amy

Tuesday, July 30, 1985

University Hospital is a strange place. I'm not really sure what I'm supposed to be doing there. I've got about eight patients, but none of them is really mine. Everyone has a private pediatrician, and the attendings are the ones who run the show. So we make rounds in the morning and decide what we'd like to do on each patient and then the attendings come around and tell us what we're *really* going to do. Having us there seems pointless. It doesn't make any sense.

We started yesterday and I got out at about three in the afternoon. Then today, I got home at three-thirty. All you do is eat lunch, write progress notes, and leave. My patients don't keep me very busy. I've got this eight-year-old named Oscar who was in a car accident or something last year. He had a really bad head injury and was in a coma for months. They had to trach him [**perform a tracheostomy: create an opening into the trachea, or windpipe**] because he was on a ventilator for three months. He's much better now, although he still needs a wheelchair, and ENT admitted him to take out his trach tube. He's been in the hospital for two days so we can watch him breathe, and he'll have to stay for another day or two. Very exciting!

My only really sick patient is this six-year-old renal transplant kid. He was born with dysplastic kidneys [**abnormally formed and nonfunctional kidneys**], and he's been in renal failure his whole life. They did the transplant yesterday morning using his mother as the donor. He was sick as a dog all last night and most of today but I couldn't even get near him! The renal service is running the whole show. Occasionally one of them will talk to me, to tell me what scut they want me to run. I probably could be learning a lot, but I can barely squeeze myself into the room!

This whole place is depressing! My patient is the newest of four transplant patients on the floor. The other three are all in some phase of rejection [**rejecting the transplanted kidneys**]. It's a real pleasure to go into their rooms; they all want to die because they know that if they continue to reject, sooner or later they're going to wind up back on dialysis.

Of my eight patients, there's only one who's anything like a regular pediatric patient. It's a three-week-old FIB [**fever in baby; all infants under two months of age who are found to have a temperature of 100.6° or greater are admitted to the hospital, have blood work and a spinal tap, and are treated with antibiotics**]. When she came in, I tried to do the workup but I had trouble getting the blood; I stuck her three times and I couldn't get a drop. I know I'm not great at drawing blood so it didn't bother me but then I called Diane Rogers [**the senior resident assigned to the**

ward that month] and she tried five or six times and couldn't get anything either. Diane got very angry, as if not getting the blood was a personal insult. Finally Dr. Windom, the baby's private, came in and tried and he couldn't get it either. We wound up just treating her as if she were septic **[had an infection in her blood]**. Windom told me to send off urine for viral cultures **[a method of determining whether a viral infection is present]**. What a waste of time and money! There's nothing you can do if it's viral, and besides, those cultures take at least two weeks. The baby will be completely better by the time we get the results back. But if that's what he wants, then that's what he'll get. I had enough trouble last month with attendings. From now on I'm just going to do whatever anybody says and not protest at all.

I was on in the ER for the last time Sunday night and it was pretty quiet. Larry took Sarah with him to his team's last softball game. They made it into the league's semifinals but they got beaten and Larry's depressed about it. I can't get over how good he is with Sarah. But Sarah is a good baby. She really has developed a personality. And she's growing like a moose! I took her to see Alan Cozza **[the chief of pediatrics at Jonas Bronck Hospital; he also is Sarah's pediatrician]** yesterday for her three-month checkup; she weighed twelve pounds, two ounces. She's gained over five pounds in three months. Alan said everything was fine. He told me we were doing a good job with her, that she seemed like a happy, contented kid. I think she is, too. I'm happy that my internship doesn't seem to be doing her too much harm.

Monday, August 5, 1985

I could take a whole year of University Hospital. It's almost like being on vacation. I've been on call twice so far and I got about six hours of sleep both nights. That's more than I usually get at home. And things have been quiet enough for Larry and Sarah to come visit me while I'm working. They spent most of Saturday afternoon at the hospital and they even had dinner with me in the cafeteria. There are almost never emergency admissions at night, and on days I'm not on call, I've been getting out by three in the afternoon.

The other two interns have been having a pretty easy time, too, but they haven't been leaving as early as I have. They stay until at least five. They don't have more work than I do, they've just been hanging around, spending time teaching their medical students and basically just looking for things to do. When I'm on call, I tell them just to sign out and go home but they won't do it. I think they feel guilty about leaving early, like it's a sign that they're goofing off. It's not goofing off; it's more like survival.

I'll tell you, even if I didn't have a baby to get home to, I wouldn't want to stay at that hospital any longer than I have to. The place is so depressing! My patient Ricky, the six-year-old boy who got his mother's kidney, stabilized last week and the renal team finally let me into his room. He's a nice kid; he's very small [the effect of his chronic renal failure], maybe the size of a three-year-old, but he's smart and he's got a good sense of humor. I talked with his mother who was discharged from the hospital on Thursday [she had been hospitalized for removal of her kidney]. She and her husband have been through hell since Ricky was born. Now, because of the transplant, they were hoping their lives might finally get back to normal again. But then over the weekend,

Ricky's BUN [blood urea nitrogen] and creatinine started to rise. [BUN and creatinine are measures of renal function. Elevations imply that the transplanted kidney may not be functioning well.] Renal decided today that he was in acute rejection; they had me order a renal ultrasound and started him on ATG [Antithymocyte globulin, a drug designed to prevent the immune system from making antibodies against the foreign kidney]. The renal attending told Ricky's mother this afternoon and she started to cry and of course then Ricky started to cry. The whole thing really upset me, so I was glad I was finished with my work. I just packed up and came home.

One of my other renal patients is a twenty-year-old with Down's syndrome [a condition caused by an extra chromosome that leads to mental retardation and other abnormalities] who also happens to have chronic renal failure. He's extremely high-functioning for a Down's patient and he understands everything that's happening. He came into the hospital last week because his BUN and creatinine were going up. He had had a cadaveric transplant [his transplanted kidney had come from a dead donor] a year ago, and they thought he was in rejection. They scheduled a biopsy [a procedure in which a needle is passed into the donor kidney and some tissue is removed; the biopsied material is analyzed under a microscope for signs of rejection] for Friday and told me to make him NPO [nothing by mouth] starting at midnight the night before. Well, of course they didn't say a word to him about the biopsy, and when the breakfast trays showed up on Friday morning and he didn't have one, he started to yell. I had to tell him they were going to do a biopsy. Nobody from the renal team had the decency even to talk to him about it!

.The biopsy was done and it did show signs of rejection, so they rolled him back to the ward and started him on ATG, too. I've got five patients on ATG now. The drug doesn't seem to cause any harm, so having all those patients on it isn't making my work any harder, but since it's only given to patients who are in the process of rejecting their transplant, there's a lot of misery attached to giving it. This ward is filled with gloom and doom.

Things with Marie have calmed down. She and I have been on good terms since I changed services last week. I think she's relieved I'm not coming home for lunch anymore. And I don't think she minds very much that I've been sending her home early two out of every three days. So, all in all, if things keep up like this, I don't think I'll mind the rest of my internship. Of course, I doubt that it'll keep up like this!

Tuesday, August 13, 1985

I was on call last night and I'm tired. It was my hardest night so far this month and I got only an hour and a half of sleep. Ricky's been really sick. He had to go back to the operating room yesterday because of complications. I really thought he was going to die.

I guess he started to go bad last Thursday night. He started complaining of belly pain. His mother told me about it when we were on rounds on Friday morning and I told her I'd check with renal but I forgot to mention it to them. I was on Friday night and he seemed to be in pretty good spirits even though his BUN and creatinine had gone up a little. He did complain that his belly was hurting a few times but whenever I examined him, I couldn't find anything wrong. He slept well and didn't get a fever or anything, so I just forgot about it. But I guess the pain got worse on Saturday afternoon and his mother told Margaret Hasson, the intern who was on call, about it and she examined him and found that he was tender all over the place and that his belly was distended. She called the renal fellow and he told Margaret to get a CBC and a sed rate and another BUN and creatinine and that he'd call ultrasound and try to arrange an emergency renal scan. Margaret said he seemed to be a little better when she went back to draw his blood, so the renal fellow wasn't as concerned when it turned out he couldn't get the scan done because there wasn't a technician available during the weekend.

Things got worse again on Sunday afternoon when the third intern, Janet, was on call. Ricky's belly became very distended again, he was complaining of more pain; now he had pain shooting down his leg. The renal fellow came in and got all the information together. He found out that Ricky's urine output had steadily dropped over the past couple of days, so he called Dr. White [**the renal attending**] at home and discussed the whole thing with him. Dr. White must have called the radiology attending at home because within an hour there was somebody there to do an ultrasound exam. They found that the kidney looked fine but that there was some problem with the ureter [**the tube connecting the kidney with the bladder**]. Dr. White thought that Ricky's ureter had detached itself from his mother's kidney and that the kidney was making urine that was slowly leaking into the abdominal cavity and causing the pain.

This was an emergency, so Janet called the urology resident and he came to see Ricky and agreed with what Dr. White had said. The resident called the urology attending at home and the urologist refused to come in! He just refused to come in; he said it wasn't such an emergency that it needed to be fixed on a Sunday night and that he'd be in the next morning to assess the situation and, by the way, that it probably would be a good idea to keep Ricky NPO and pre-op him [**do everything necessary for surgery**]. By this point Ricky was in intense pain and his temperature went up to 102, so Dr. White was called and he went nuts! He called the urologist at home and they yelled at each other for a while, but the result was still the same, the guy wasn't going to come in until the next day. Since there was nothing he could do about it, Dr. White called Janet, told her to start Ricky on broad-spectrum coverage [**antibiotics to cover a wide range of possible bacteria**] and pain medication. It was terrible; his mother and Janet and Ricky's nurse were up all night with him.

Then finally, yesterday morning, the urology attending showed up. By the time we started work rounds on the floor at eight o'clock, Ricky was on his way to the operating room. I saw him before he left; he looked terrible. His belly and his right leg were tense and swollen. It turned out he had so much urine in his abdomen that some of it had worked its way down into his leg. It was a real mess. He was in the OR four or five hours. First they had to clean out all that urine. Then they had to reattach the ureter to the kidney and make sure it was working okay. He spent another three hours in the recovery room and he didn't get back to the ward until after four in the afternoon. The renal fellow and I were with him all night. Plus I had four new admissions to work up, nothing serious, just more pre-ops for tomorrow. It was far and away the worst night for me at University Hospital, and now I'm exhausted. I couldn't even eat dinner. Larry fed the baby and took her out for a walk and I've just been lying here, not able to get out of bed.

I don't think I've ever seen anybody as angry as Ricky's mother. Most of the parents at University Hospital are weird. They're very private, and they protect their privacy and that of their children. I guess it's understandable, since so many of them are chronically sick and wind up spending so much time in the hospital with all these doctors and nurses and medical students and other people constantly going in and out of their rooms. So they give you a really hard time when you have to do something and it makes you feel as if you're intruding into their space. Ricky's mother is different. She's friendly and she likes to have company in the room. But she's been seething since she found out the urologist refused to come in Sunday night. She's sure the delay in getting Ricky to the operating room is going to cause permanent damage to the kidney, her kidney, the only kidney she can give to her son. Dr. White told her he didn't know what effect this would have, that only after everything's back to normal will we be able to figure out what's what. She was still livid when I left the hospital this afternoon. I hope everything works out all right. But in the meantime, Ricky's on about a hundred medications, he's still NPO, and he's still really sick. There is at least one good sign: His BUN and creatinine from this afternoon were down.

I can't stay awake any longer. I'm going to sleep.

Monday, August 19, 1985

It's been a good week. I've been getting sleep, I've been coming home early, I've been spending time with Sarah and Larry. I can't complain.

The other interns on the team seem depressed, and I can understand why. There just aren't any normal children on this ward. Very few of these patients wind up being normal at any time during their lives. There's almost no hope here. On rounds yesterday, Janet said there were two kinds of patients at University Hospital: the ones who cry when you stick them for blood and the ones who don't. She said the ones who cry are bad because they make you feel guilty when you stick them. But the ones who don't cry are much worse because they're the ones who know that crying isn't going to do them any good.

I know what she meant and I know that if I were to hang around the hospital as much as she and Margaret do, I'd be depressed, too. But I can escape; I've got Sarah to run home to. And that makes everything a lot better.

Here's an example of what we have coming into this place: When I was on last Thursday, I got a five-year-old girl with intractable seizures who was coming in to have her anti-convulsant medication manipulated. She had metachromatic leukodystrophy [**a rare inborn error of metabolism caused by deficiency of the enzyme arylsulfatase; the disease leads to severe neurological abnormalities**]. Metachromatic leukodystrophy: You know how many of those there are in the world? Maybe a dozen. And one of them comes waltzing into University Hospital to have her seizure medication changed. She's still on the ward. She's on five separate anticonvulsants and we're raising one and lowering another. So far I haven't seen any real change in her. She could wind up staying in the hospital for months.

Her story's frightening. She was completely normal for the first year and a half of her life, and then her mother noticed she was getting clumsy. She started falling down a lot and losing her balance. She took her from doctor to doctor until she saw Dr. Rustin [**a pediatric neurology attending**] who made the diagnosis immediately. And now she's a GORK [**an acronym for "God only really knows"**] with an intractable seizure disorder.

The first thing I thought about when the mother told me the story was, Can something like that happen to Sarah? I mean, she seems completely normal now, but who knows what might happen next year or next month or next week? Who knows? Thinking about things like this can drive me crazy! I find myself doing it a lot. Every time I admit some-one with something strange I think, Can Sarah get this? It usually doesn't stay with me for long, I can shake it pretty fast, but when it happens, it's like a wave of terror passing through me. I don't even want to think about it.

Ricky's much better. He had a pretty rough time last week but by Friday he was about back to normal. His kidney seems to be functioning well, and his BUN and creatinine came down to all-time lows today. He's been out of bed and walking around the ward, playing with some of the other patients. His mother's also calmed down a lot. Dr. White is pretty sure that no harm was done to the kidney but he says we'll still have to see what the future holds. He plans to send Ricky home either tomorrow or the next day.

Sunday, August 25, 1985

Today is Sarah's four-month birthday and we had a little party. My father came and so did Larry's parents. It was the first time we'd all been together since Sarah was born. My father was doing pretty well; he looked good and he seemed happy. Everyone was worried about me. They thought I looked pale and tired. I should look pale and tired; I'm work-ing hard and I haven't exactly had a lot of time to go sunbathing, but I told them I think I'm doing okay. I think I am. I think I'm doing better than most of the other interns I've seen around.

Sarah rolled over for the first time last week. She did it first for Marie. Marie said she put her down in the crib on her belly and when she came back a few minutes later, she was lying on her back. So she put her back on her belly again and watched her, and sure enough, she flipped right over again.

I've been getting along very well with Marie. She really does love Sarah. I think she's been holding back on the feedings a little and not carrying her around as much. At least that's what she's been telling me. So things are going well on that front.

The renal team discharged Ricky on Friday. The nurses had a little going-away party for him. It was really nice. He was definitely my favorite patient of the month.

I finish this rotation on Tuesday. I'm on tomorrow night, the last night of the month, and then I start on 8 West [**one of the general pediatric wards at Jonas Bronck**]. Going to 8 West'll be like coming home. I did my subinternship and my third-year rotation there. I'm looking forward to it. But I know I'll never beat the hours I've been able to keep at University Hospital.

Mark

Sunday, August 4, 1985

I started on Infants' [a ward at Mount Scopus Hospital] last Monday and so far this place makes Children's look like an amusement park! I was on yesterday; I worked my ass off all day long, running from one thing another; and at no time did I have any idea what the hell I was supposed to be doing. Usually, when you're on call on the weekend, you start with work rounds where you and the resident decide what needs to be done on each patient. It doesn't work quite that way on Infants'. First of all, when I got to work at eight o'clock, the resident who gave us sign-out was a cross-coverer [a resident who works in another part of the hospital during the day and covers the particular ward at night only], and she didn't have much of an idea of what was going on with the patients. We didn't get any kind of intelligible sign-out, so we started off with one strike against us. And then when we finally got everything sorted out and came up with a plan of what we wanted to do for each kid, the private attendings starting calling to tell us what actually was going to be done. And then there were all these admissions coming in. I just wanted to say, "Okay, I'm going to go outside now and come back, and then we'll start the whole damned day over again."

I picked up some real terrific patients when I came over to Infants'. I've got this incredible specimen named Hanson, who's four months old and has never been out of the hospital. When we went into his room on Monday morning, he was lying there in his crib, weighing all of about two pounds, with these wasted, shriveled arms and legs that were stiff as boards. He wasn't able to suck on a pacifier, and it seemed like he was having these little seizures. He looked like warmed-over death, and the senior actually said he was looking good that day compared to how he looked last week. My God, he must have looked like rotting hamburger the week before! It turned out he had crumped before we changed services and when they worked him up [in this case, the workup consisted of blood cultures, a spinal tap, and urine cultures] they found he had a disseminated fungal infection. A fungal infection! Now, there's a common cause of a crump. But I guess it wasn't so strange in this kid: He's had chronic diarrhea for the past two months and he hasn't gained an ounce in all that time. Since his mother's an IVDA [intravenous drug abuser], we're sure he'll be a candidate for admission to the AIDS clinic.

Anyway, he's being treated with amphotericin [a drug for systemic fungal infection], which is so toxic that even if the infection doesn't kill him, there's a good chance the treatment will. It's got to be given by IV. He has a central line [an indwelling catheter passed through the skin into one of the major veins in the chest], but we're giving him his TPN [total parenteral nutrition, a treatment in which a large number of calories are provided by vein] through that, so we have to give him the amphotericin through peripheral lines [normal IVs]. His IVs usually last only about twenty minutes, and most of the veins in his arms and legs are already blown, so I can see this kid is going to take up a lot of my precious time this month.

The people on my team seem pretty good. There's Elizabeth, of course. She already told me she doesn't like Infants' and that if it were all the same to the chiefs, she'd rather be back on Children's. And then there's the other intern, Valerie Saunders. I don't know about her, she seems kind of depressed. Our resident is Rhonda Bennett. She's smart, but she treat us like we're real morons. I mean, on rounds in the morning, she makes sure to go over every little detail two or three times, and then makes us repeat what she says and write it all down. It's like being in first grade or something. Elizabeth said something to her like, "C'mon, Rhonda, we promise we won't forget, cross our hearts and hope to die," and she got real defensive and said she was just trying to help us and make it easy for us. Well, I'll tell you, if she keeps it easy for us, I might have to murder her.

Wednesday, August 7, 1985

I'm going to kill them, I'm going to kill them all! I was on last night and today was the worst day of my internship. It's bad enough spending the night running from room to room trying to keep twenty-eight babies from dying, but to do that and to have to spend the next day being nice to Rhonda *and* putting up with all the shit the chief residents are hand-ing us, that's a little too much. So it looks like I'm going to have to kill everybody to get any peace.

The first one I'm going to kill is that Hanson. He crumped again last night. He stooled out [**developed diarrhea**] and got acidotic [**built up acid in his blood, a sign of deterioration**] and shocky. I had to do a whole sepsis workup including a spinal tap and pull out his old IV and start a new one; the whole thing took over four hours. And then I had to call the ID fellow [**the fellow covering the infectious disease service**] and argue with him about what antibiotics to start him on. He told me to use three drugs, two of which I'd never even heard of before!

The second one I'm going after is Rhonda. She's so damned cheerful all the time, it's disgusting! At two o'clock in the morning, after I got off the phone with the ID fellow, I went to tell her what antibiotics he had suggested and she smiled and said, "Well I don't know about that, Mark, I don't know if those antibiotics give adequate coverage against enteric gram negatives [**bacteria that normally inhabit the intestinal tract**]. You did tell him that Hanson had chronic diarrhea, didn't you?" Of course I hadn't mentioned the kid's diarrhea. It was late and the kid had been trying to die on me all night and I can't be expected to think of everything! So, still smiling, she ordered me to call the ID fellow back and rediscuss the case with him. Of course the guy knew Hanson had diarrhea; he had suggested the drugs just for that reason.

The thing about Rhonda is if you go and tell her she's a pain in the ass and that she's making your life miserable, she takes it personally and starts to get teary-eyed. So even though she is a pain in the ass who's making my life miserable, I have to be nice to her anyway. I don't think I can take this for a whole month. So I'm pretty sure I'm going to have to kill her.

And the third one I'm going to have to kill is Arlene, the chief resident. There I was, sitting in the residents' room at noon today, minding my own business, trying to catch my breath; I'd made it through the night; I'd worked up six admissions. I had managed to keep Hanson and all the rest of them alive; I had even managed to make it through work rounds and attending rounds without falling asleep or complaining much. All I wanted to do was finish my scut, write my progress notes, and get my ass out of there. But could I do that? No! Arlene came in, saw us interns sitting there, and she said, "Aren't you guys going to the noon conference?" Well, Elizabeth said she had to start an IV on a kid who was supposed to go to the OR at one and Valerie said she had something else to do, and I just sat there unable to move. So Arlene said, "You know, these conferences are for you guys, not for us. It's just more work for me to schedule them. If you interns don't want to come to them, maybe we shouldn't schedule them anymore." None of us said anything back to her. I just glared. Here I was, having killed myself all night, having killed myself for over a month now. Maybe you'd think the chief resident ought to come up to us and compliment us every once in a while, tell us we're doing a good job and that we should keep it up, but no, all we get told is that if we don't come to conferences, they're going to cut them out! So if she ever says anything like that to me again, I'm definitely going to kill her.

I'm worried about all of us, but I think Val's in a lot more trouble than Elizabeth or me. She's really depressed. She says she'd rather be hiding under her bed than working in the hospital. Now I'm no psychiatrist, but that sounds pretty abnormal to me. She was on Sunday and spent the whole day trying to start an IV in Hanson and doing a lot of other technical scut. To hear her tell it, she missed every single time. And then the senior resident would come along and plop a needle in and get it on the first stick. Val got so frustrated that by Monday morning she couldn't even get blood from the veins of the easy kids. She walked around like a zombie most of the day. Rhonda had to tell her to go home in the early afternoon because she wasn't doing anybody any good. I think Rhonda felt better about Val leaving. I get the feeling Rhonda would be happiest if we all would leave. That way, she'd just take care of all the patients herself without anybody to bother her. Have I mentioned yet that I'm going to have to kill her?

Well, all this may not make much sense, but it sure as hell made me feel better to get it off my chest. I can now go to sleep without worrying about tearing my pillow to shreds.

Thursday, August 8, 1985

Maybe Rhonda isn't so bad after all. At the end of attending rounds today, Claire, the other chief resident, came into the residents' room and said, "It's come to my attention that maybe we haven't been paying enough attention to you guys." That's an understatement! She told us how sorry she was about it and that she wanted to find out what the chiefs could do to make our lives easier. And before anyone could say anything else, Rhonda yelled, "This makes me so damned mad!" and immediately broke into tears. She caught her breath and said, "Here we are, working our rumps off. I had eleven admissions the other night [on **nights on call, Rhonda was responsible for all patients** admitted to both the Infants' and the Children's ward] and Arlene knew it but not once did I get a 'You did a good job last night, Rhonda' or anything. All she gave me was, 'If you can't get your interns to conferences, we just won't have them anymore.' "

Then Claire got a real concerned look on her face and asked, "Rhonda, what's wrong?" and Rhonda yelled back, "You want to know what's wrong? You treat us like dirt! It wasn't so long ago that you were doing this! You can't tell me you don't remember what it's like to be the senior on Infants' with all these sick kids and all these admissions and all the attendings coming around to bombard you with demands every second of the day! But neither of you seem very sympathetic. All you can do is complain that we're not coming to conferences. You know I'd love to be able to go to the conferences, I'd like to learn something. But I don't see you or Arlene volunteering to cover the ward for me so I can go!"

I wouldn't have believed it if I hadn't seen it. Elizabeth felt the same way. Neither of us thought Rhonda had it in her to stand up for herself like that. She seems like too much of a robot to show that much emotion. I mean, she's feeling as rotten about working on this ward as we are.

The rest of the exchange was pretty amazing, too. After Rhonda finished yelling, Claire said, "Rhonda, you know what we think of you. We might not always say it, but you're the best we've got. Whenever I see your name on the schedule, I breathe a sigh of relief because I know you're never going to do the wrong thing." And then Rhonda said, "You sure have a strange way of showing it. I don't expect a pat on the head just for taking night call, but I don't expect to be yelled at either." They talked a while longer after that, but it wasn't as good as this first part. It was pretty amazing. It made me feel a little better about working with Rhonda. Who knows? Maybe I won't have to kill her after all.

Tuesday, August 13, 1985

What a calm, relaxing night last night was! I got four admissions, all of them in the middle of the night, all real simple: a kid with congenital heart disease that's so complicated I need a medical dictionary, an anatomy textbook, and a road map just to get through the old chart; I also got a ten-month-old who had GE reflux **[gastroesophageal reflux, the reflux of acidic stomach contents back into the esophagus]** that was corrected surgically when he was a couple of months old, who got diarrhea over the weekend and got himself pretty dehydrated; a straightforward meningitic who happened to be seizing; and a three-year-old with meningomyelocoele **[a congenital defect of the spine that causes paralysis of the legs, bowel and bladder incontinence, and hydrocephalus; also called spina bifida]** who came in with a high fever and looked like shit. We thought he probably had meningitis, too, but it turned out he probably only has a UTI **[urinary tract infection, a common problem in children with bladder incontinence]**. I managed not to get any sleep again. And then today my pal Hanson, who was getting better, decided to get a fever. He looked pretty good so I didn't make too much out of it. I figured he had the virus that's going around but then Rhonda heard about it and took a look at him and said, "Well, it may be the virus, but I don't like the look of those IVs" **[fever in a child with IVs can be caused by infection of those IVs]**. I was planning to spend the day writing my notes and getting the hell out of there. But did I do that? Of course not! I wound up spending the afternoon sticking needles into Hanson's body, trying to start new IVs. I must have stuck him ten times before I got one in. The kid's totally aveinic [**internese for "without veins"**].

My new diarrhea patient has a strange story. He came in with his grandmother, who said he got all his care at another hospital but she doesn't remember the name of either the hospital or the doctor. She said she came to Mount Scopus this time because that other place had the kid for all those months and they couldn't do anything to make him better, so she was coming to give us a chance to cure him. To tell the truth, he didn't look that bad to me, but to hear his grandmother tell it, he's at death's door. I'm going to have to figure out what's going on with him, but I sure as hell wasn't going to do it today.

So finally I sat down to write my notes and got out of the hospital at about four-thirty. My progress notes have gotten worse and worse. It's gotten to the point now where I can't even read my own handwriting. An attending came up to me yesterday and asked me what I had written on his patient's chart and I simply could not read the thing. I'm pretty sure I'm going to get yelled at about my handwriting sooner or later. But what can I do? If I decided to take my time and write neatly, I'd never make it back to my apartment. It's kind of a shortcut I've got to take to keep my sanity at this point. Maybe this is how the doctors' handwriting myth began.

Monday, August 19, 1985

Things are looking up. Really! Last night wasn't bad, I only got one hit [**hit=admission**], and for the first time this month I actually got into the bed in the on-call room and fell asleep for a while. And Hanson is better. His fever went away without any change in his antibiotics, so either it was the virus that was going around or maybe one of his IVs actually was infected. We started feeding him formula again last week- end [**he had been NPO for a few days following his most recent episode of diarrhea**], and he's tolerating it pretty well. He hasn't had any diarrhea and he actually gained a few ounces. He's a pretty cute kid, actually. I'm getting to the point where I actually like him. If he behaves himself and doesn't crump or do anything stupid like that, he may become one of my favorite patients. We're even starting to think about sending him home. The only problem is, his mother, who's an IVDA, has never come to see him. I've never met her or even spoken to her on the phone. So it looks like he's going to turn into a social hold. I've got to start talk- ing to the social worker about placing him somewhere. Oh, well, he'll probably wind up staying on Infants' until I'm a senior resident.

And that patient with the meningomyelocoele I admitted last week turned out to be a great kid. It's a funny thing about him, he turned out to be kind of cute. He'd sit in his little stroller and make this weird clicking sound with his cheek to get your attention, and when you'd look over at him, he'd smile at you. I liked that kid a lot and I really miss him since he went home. He was the only kid I've taken care of this month who's old enough to actually be sociable.

So far, the weirdest story of the month has to do with Fenton, that GE reflux kid I admitted on Monday. I sat down and talked to his grandmother on Wednesday. She's the kid's caretaker; his mother's about fourteen and is treated more like an older sister. Anyway, the grandmother told me this real bizarre story. She said he vomited everything they fed him when he was a little baby and she brought him to some hospital in Westchester, which we all finally figured out had to be Westchester County Medical Center [**a teaching hospital affiliated with New York Medical College, in Valhalla, New York**]. They worked him up, diagnosed the reflux, and did a fundoplication and a feeding gastrostomy [**placing of a tube directly through the abdominal wall and into the stomach, to facilitate feeding while the esophagus is healing**]. But he never seemed to get any better after the operation. The grandmother took him home but he kept vomiting whenever they fed him anything by mouth and got diarrhea when they gave him anything through the g-tube [**gastrostomy tube**]. She kept bringing him back to the hospital and they finally started him on continuous gastrostomy drip feedings [**sort of like an IV, delivering small amounts of fluid throughout the day and night, into the stomach**]. She told me that that was the only thing that seemed to work.

Well, none of this made any sense to any of us, including Dr. Gordon [**the pediatric gastroenterologist**]. There's no way this kid could have so many problems and look so healthy. And the grandmother is a pretty suspicious character; she knows all the medical terms and the names of all the procedures. So yesterday we called the gastroenterologist at Westchester County Medical Center and he told us what really was going on. He said that the grandmother kept bringing him to the ER there with a history of diarrhea and vomiting but the kid never looked dehydrated. They admitted him a few times and he did have loose stools but for a long time they couldn't figure out what was happening. Finally, during an admission about a month ago, one of the nurses found a bottle of laxative in the grandmother's possession. They couldn't prove it, but they're convinced she was giving the kid the laxative in his bottle to make him have diarrhea. Amazing!

So today, while the grandmother was off the ward, we started the kid on regular feeds and he took it like a normal child. When the grandmother showed up, she got really angry and tried to sign him out of the hospital AMA [**against medical advice**], but we stopped her and slapped a BCW hold on the kid [**the Bureau of Child Welfare can order a child retained in the hospital if the child's well-being is endangered**]. The grandmother went crazy but the social worker talked her down; the social worker handled the whole situation pretty damned well.

Well, there's only about another week of this insanity left. I can't wait. I've had about enough of this Infants' nonsense!

Friday, August 23, 1985

I meant to record this yesterday, but I fell asleep as soon as I got home and I couldn't do it. Wednesday was another classic night on Infants'. I'm beginning to lose my sense of humor about all this, which is a pretty serious problem. It's definitely time to get off this ward. I'm going to OPD [**Outpatient Department—the ER and Clinics**] for two weeks and then I've got vacation.

Well, Hanson crumped again yesterday morning. He started stooling out again and got acidotic, and while we were trying to start an IV his heart rate dropped and we had to call a CAC [**resuscitation for cardiac arrest**]. We got him back but the chiefs decided he was sick enough to be transferred to the ICU, so we shipped him up to the sixth floor. Just like that! I don't know, he's fine as long as he doesn't do anything to bother you. But the kid crumps at least once a week! He's got to learn a lesson if he expects anyone ever to like him.

And Fenton is fine, absolutely fine. His grandmother has become a basket case, though; she simply can't cope with the fact that he has no medical problem. It's really weird. The grandmother told one of the nurses that she herself has had over twenty operations; she even had a CAT scan last week while the baby was in the hospital because she's afraid she's got a brain tumor. The nurse pointed out to us that she wears one of those plastic hospital bracelets as jewelry! The social worker has been trying to get her into some sort of therapy but the woman is resistant. I'm not sure, but I think it's going to come down to either the woman gets some form of help or the baby is going to be placed in a foster home.

I'm on tomorrow for the last time on Infants'. I can't wait to get it over with. Carole and I are going to go out for dinner Sunday night to celebrate. I'm really afraid I won't find anything funny anymore. I really think I've lost my sense of humor on Infants'.

Bob

Although I was a medical student at Schweitzer and did my residency at Jonas Bronck and the Schweitzer University Hospital, I was an intern at a medical center in Boston. I left the Bronx because it was suggested that I should see how medicine was handled at places other than those associated with the Albert Schweitzer School of Medicine. So I spent a year in Boston; I'm still recovering from it.

I did my first month of internship in the neonatal intensive-care unit of a maternity hospital that was affiliated with the program's main teaching hospital. I arrived at work on the first day, a Saturday, and took sign-out from the old intern who had been on call the night before. After he left for home that morning, I was pretty much left on my own with thirty-five of the sickest premature babies you could possibly imagine. That first day of internship was definitely in the top ten of the most frightening days of my life.

When I started in that NICU, I knew absolutely nothing; the intern who signed out to me communicated in what seemed to be a foreign language. He spoke a hodgepodge of medical terms, slang, and numbers all mixed together. I just wasn't ready for: "That's a forty-five-hundred-gram IDM who aspirated mec and got PFC. We tubed him and put him on the vent with settings of twenty-five over five, 100 percent, and forty, and his last gas was seven point thirty, forty-four, and forty-five. He blew two pneumos so we put in tubes. He's on DIOW at eighty per kilo per day." I had absolutely no idea what any of this meant; I just wrote as much of it as I could on my clipboard, nodded my head to make him think I understood what he was saying, and hoped to God that the nurses knew what the hell was going on.

I eventually figured it all out. It didn't take long before I could translate even the most complex of these monologues into English. (By the way, the intern was talking about a nearly ten-pound newborn whose mother was a diabetic. The baby had passed a bowel movement while still in the womb and had breathed in the contents of the bowel movement, causing severe respiratory and cardiac problems. He was being breathed for by machine, had too much acid and not enough oxygen in his blood, had had two episodes of collapsed lung, and was being given intravenous sugar water. That baby, one of my first patients, survived and did fairly well in spite of me.) And eventually I even became comfortable with the preemies. But that Saturday was terrifying for both me and my patients.

I spent August back at the medical center, working in their NICU. Although I was feeling more comfortable with preemies after my month at the maternity hospital, I encountered many other problems. First, two months in a row in a NICU is cruel and unusual punishment. Preemies, unlike older children and adults, don't seem to understand the difference between day and night. They didn't discriminate: they'd crump at any moment, morning, afternoon, evening, and in the middle of the night. As a result, when working in a NICU, it's almost impossible to get any sleep during nights on call; you usually don't even get a chance to see the inside of the interns' on-call room.

Second, although the neonatologists will tell you that saving preemies is an exciting and exhilarating experience, to me the unit was an unbelievably depressing place to work. There were a lot of deaths, and although dealing with the parents of the babies who died was sad and difficult, it was even harder to care for some of the very tiny and extremely sick infants who didn't die. These survivors often didn't have a snowball's chance in hell of leading anything resembling a normal life. Yet we were ordered to do everything possible to keep them going, and their parents were often given unrealistic expectations about how their infant would turn out. That conflict between what was medically demanded and what seemed ethically correct took a toll on me and on a number of my fellow interns.

The third problem that struck me when I made it back to the medical center was that I felt alone. There were two reasons for this. First, all the other interns had met and become friends during July. By being farmed out to the maternity hospital, I had become "odd man out." It took me months to make inroads into the cliques that had formed.

The other reason I felt alone was because my wife and the rest of my family were back in New York. In a situation almost parallel to Andy Baron's, while I was off in Boston, my wife was a graduate student in New York. We would see each other only on those weekends when I had at least a full day off. Since that happened only two of every three weeks, there were long stretches of time when I was completely alone. Without friends and family, my life was miserable.

And miserable was the tone set for the entire year. I felt overworked, dead tired, conflicted by what I was being called on to do, and uncared for by the senior people in the program. And even though I had originally planned to stay in Boston for the three years of my training, I decided to leave the medical center after my internship. I made my first call to Alan Cozza, the chief of service at Jonas Bronck Hospital, asking for a job as a junior resident toward the end of August. By September I informed my chief resident in Boston that come the following July 1, I'd be moving back to the Bronx.

In retrospect, my experiences in Boston were not unique. All interns suffer during their internships. Although there might be some variations, the issues are pretty much the same for everyone. The main issue is the hours: Being on call every third night all year long makes it impossible to lead anything like a normal life. Regardless of how caring the people who run the program are, or how nice the city in which it's placed is, or how much support is available from family and friends, interns usually spend a hundred hours or more per week in the hospital. And anytime someone spends that much time at their place of work, there are going to be problems.

But why do house officers have to spend so much time in the hospital? What do interns do all day long? To explain this, I should outline what a typical intern's day is like.

On a typical day, most of the interns show up for work at about 7:30 A.M. They briefly walk around the ward, making sure that all their patients have literally survived the night. They check the vital-sign records kept by the nurses to see if the patients have had fevers or any other complications. Then, at about eight, work rounds begin. The ward team, made up of three interns, a resident, the head nurse, and the third-year medical students who are assigned to pediatrics that month; walks past each patient, reviews his or her progress and decides on a plan of action for the day. The interns must carefully note the plans for each of their patients; it is their job to make sure the plans are carried out, to order the tests, schedule the appointments, send off the lab specimens, and check on their results. It is at work rounds, which last until approximately nine o'clock, that the interns generate the "scut lists" that will occupy them for most of the rest of the day.

At nine, an intake conference occurs. At intake, all patients admitted the night before are reviewed with the chief of the service. This is a teaching conference, and a large portion of the house staff usually is present. The interns are expected to present their own patients briefly and, if recommendations regarding management are made, to add these to their usually already burgeoning scut lists.

Intake lasts until about nine-thirty, at which time an X-ray conference begins. At this conference, all X rays taken the day before are reviewed with the radiologists. This X-ray conference usually lasts until ten o'clock.

Then comes attending rounds, when the ward team meets with a member of the faculty. During attending rounds, admissions from the night before are focused upon, the presenting symptoms dissected, and the patients' diagnoses discussed at length. The ward attending is the person who is ultimately responsible for the care that's delivered, and so in addition to teaching about the conditions that afflict the patients, the attending must make sure that the proper things are being done in a timely fashion. Depending on how many patients were admitted the day before and how long-winded the attending is, rounds can go on until between eleven o'clock and noon. Every day at noon there is a didactic lecture on a pertinent topic in pediatric medicine. So, the average intern may not get down to attacking the scut list until after one o'clock in the afternoon.

Most interns will tell you that scut is the sole reason for their existence. Scut includes blood drawing, IV starting, the tracking down of lab results, the ordering of diagnostic tests, the calling of consulting services, and finally the writing of progress notes. Most of this stuff is sheer frustration and takes hours and hours to complete. While "running the scut" the intern also is responsible for teaching his medical student about pediatrics. Depending on how many patients he's following, how efficient he is, and how many questions his medical student asks, the intern who's not on call may get out of the hospital anywhere between three in the afternoon and nine o'clock at night, with the average being around six.

When they're on call, of course, they don't go anywhere. They stay all night, managing any complication that may arise in any of the patients on the ward and admitting all new patients who are sent up from the emergency room. Sometimes, when the emergency room is quiet and the patients on the ward are stable, the intern might be able to retire to the on-call room to get some sleep; at other times, when things are hectic, he or she might not even have enough free time to go to the bathroom. And the daily routine begins again at seven-thirty the next morning; even if the intern has gotten no sleep during a night on call, he or she is expected to participate in all the activities that occur during the entire post-call day.

This cycle is repeated every third night. Interns spend the first night in the hospital. The next night, when they're postcall, they usually are unable to do anything other than go home and hit the sack. The final night in the cycle, the precall night, is the only one in which most interns feel alive enough to go out and have a little fun. But very often, the precall night is ruined by anxiety; lurking in the back of the intern's mind when they're precall is the fact that the following night may be a complete and utter disaster. And so, in a sense, even when they're out of the hospital, there's no escape.

The interns are also expected to carry out certain tasks that are not all that difficult when well rested but may prove to be impossible after a night spent on call without any sleep. Without sleep, an intern can lose track of the subtle social skills that are necessary for communication; as a result, talking to patients and their families can become torture. The intern also is expected to present orally, during attending rounds the next day, all the patients who were admitted during his or her shift. Keeping track of names, symptoms, physical findings, lab results, and treatments can become an insurmountable task when you're having trouble just keeping awake. And screwing up a presentation can bring on the wrath of the attending, who is relying on the intern's information, and a lowering of the intern's own self-esteem.

This system of night call has come under a great deal of scrutiny in recent years. Public awareness, however, has not been focused on the toll that these long shifts are taking on the interns and residents, but rather on the toll that they're taking on the patients. It's been argued that a house officer who's been up all night can't possibly provide adequate care for critically ill patients. So, over the past few months, some alternatives to the current system have been proposed. The most popular of these would limit both the number of hours an intern or resident could work during a single stretch to twenty-four, and the number of hours worked within a single week to eighty.

On the surface, this seems as if it would be a good situation, but some house staff members have expressed fear that new regulations such as these would actually make their lives more miserable. These house officers recognize the fact that to provide staffing of the wards and emergency rooms on a twenty-four-hour basis, hospitals would have two choices: Either hire 25 percent more interns, or have the existing interns work twelve-hour shifts seven days a week. Because of the lack of availability of funds to pay for a whole crop of new house officers, as well as the problem of finding qualified medical school graduates to fill these positions, people are worried that the second choice is the one that would be instituted. And almost everyone agrees that they'd much rather work thirty-six hours at a stretch and have a day off every week than work shorter periods every day of the week.

The only hospital in which the intern's day deviates significantly from what I've outlined above is University Hospital. University is a hospital with a split personality. On the one hand, it seems like a laid-back, friendly community hospital nestled in a neighborhood of two-family houses; all the patients have private attendings (in sharp contrast to the two municipal hospitals, where the opposite is true), the nurses and the rest of the staff are like the boy and girl next door, and the pace is slow and relaxed. This makes University Hospital seem like a place you might actually like to visit during your summer vacation. However, the hospital is a major teaching affiliate of the Schweitzer Medical School and therefore is in reality a high-powered academic center. It's the place where many of the full-time clinical faculty of Schweitzer admit their "interesting cases" for special studies. As such, the hospital contains patients with rare and often deadly diseases who need vigorous, round-the-clock management. Trying to fit these two personalities into the same building is not the easiest job in the world. And who suffers because of this? The interns and residents, as usual.

There are very few teaching conferences and much more free time at University than at the other hospitals. During the day, the interns work as glorified secretaries. Each patient's attending really makes all the important decisions affecting the patient's care, but attendings are not permitted to write in the nurses' order book. The nurses are instructed to pick up and carry out only those orders written by an authentic intern. So any time an attending wants to change a medication or order a test, he or she must get hold of an intern and ask that an order be written. The interns rarely have the opportunity to argue with an attending's request. They simply have to write down exactly what's been dictated.

Although there are very few emergency admissions, interns on call frequently spend a good part of their night fighting with the lab technicians. University's community hospital personality carries over to its laboratory. At night there are very few technicians covering the hematology lab, the chemistry lab, the bacteriology lab, and the blood bank. Because of the shortage of personnel, the technicians are never exceedingly happy about running any tests in the middle of the night, and if an exotic test needs to be done, they can turn downright ugly! Since a fair number of the patients on the ward can be very sick, it sometimes becomes critically important to get tests done after midnight. And this often results in massive arguments.

The patients at University Hospital are exceptional, to say the least. One reason people who train in pediatrics are attracted to the field is because children are basically healthy; their recovery usually is rapid, and it's a rewarding experience for the doctor. But at University Hospital, you have a ward full of children with uncorrectable chronic diseases. The pediatric renal service is housed at University Hospital, and all the kidney transplants are performed there. At any one time the ward will have five or so kids whose kidneys don't work and who are either waiting for, in the midst of recovering from, or actively rejecting a renal transplant. Except for the patients who have recently gotten a new kidney, few of these children are acutely sick. That's a mixed blessing: The chronically sick patients don't require a great deal of concentrated hard work, but they usually don't get remarkably better. And that can be discouraging.

Interns find different ways of coping with the aggravating parts of working at University Hospital. Some get into fights with the patients and staff; some spend their time hanging out in the cafeteria; and some try to get out of there as early as possible. The interns spend only one month out of their year there, so the rotation doesn't usually cause any serious or long-lasting damage.

Andy

Friday, August 30, 1985

I'm out of the NICU; I made it, although I had some question about whether I would that last night when I had to supervise at the death of a full-term kid who had aspirated meconium [**meconium is the first bowel movement; fetuses who are stressed intrauterinely frequently pass meconium before birth; they then breath it into their lungs during their first inspiration and develop a severe meconium pneumonia as a result**] and who wound up on maximum doses of tolazoline, dopamine, and the highest respirator settings possible. He finally died at four in the morning. The rest of the month is all a delirious blur. I think I actually learned something, but I just don't know whether it was worth the price I had to pay.

And now I'm on the Adolescent floor at Mount Scopus. I've been told it's easy street, but I don't totally agree. Life is certainly better, though. The veins of these kids look like pipelines. No more four hours wasted trying to start an IV. So far I haven't been beaten or abused, and if I can get a good night's sleep tonight, I'll be rejuvenated for call tomorrow. You know, I've been so burned out lately, I just hope that maybe in the next few days I'll get excited and interested again.

I almost thought I was getting excited and interested today. We had rounds with Marilyn Connors, our attending. She was pretty laid back. I've heard that all the adolescent attendings are hyper and picayune, so we'll see.

There's this intern I'm working with, Margaret Hasson. She was hysterical today, the way you get sometimes when you're postcall and can barely stay awake. She was presenting a patient she admitted last night and she got really out of line, making off-color jokes and stuff. When you get so tired you're falling asleep on your feet like that, you think things are hysterically funny when nobody else does. It was great. After rounds, she told me that after she spent July over on 8 West, she decided she totally hated her internship. She doesn't seem like she's in a bad mood now, though. She's not demoralized or grumpy and she doesn't hate everything.

I'm having my first experience with teaching medical students this month. I was assigned a good student who's very conscientious, humble, and a hard worker. I try to get her to spend as little time as possible in the hospital. She should be home, reading. That's what all medical students should be doing. They should spend only enough time on the wards to get an idea of what goes on there. I didn't do that; I think I spent too much time on the wards. Maybe that's why I'm so burned out already.

Having a student is interesting. I'm finding I do know a few things. I didn't think I had picked up anything since I got

here, that I was a complete dum-dum. But I'm finding that I
can actually talk about subjects intelligently. I don't know if
they're sitting there thinking, Gee, this guy's really stupid,
he's saying things that are completely wrong. But the important thing is I actually learned something over the past two
months. I don't remember it happening; it must have been
by osmosis.

We've got these two residents, Nancy Rodriguez and
Terry Tanner, working on Adolescent this month. Nancy and
Terry seem real nice, but they're only second years; they were
just interns a couple of months ago. They're only a few
months ahead of me and they're supervising on a busy ward!
It's scary. I couldn't do it.

Saturday, September 7, 1985

This is supposed to be an easy ward, but God, there's just
so many goddamned frustrations! Like there was this patient
admitted the other day, this cute little fourteen-year-old girl
from Barbados with severe mitral stenosis [**a tight closure of
the mitral valve, the valve between the left atrium and the
left ventricle of the heart; stenosis results in the atrium
having to work harder to push an adequate amount of
blood into the left ventricle and to the rest of the body**]
as well as pulmonary hypertension [**an irreversible increase
in the pressure in the blood vessels that carry blood from
the heart to the lungs; frequently the cause of death in
children with congenital heart disease**] and hemoptysis
[**coughing up of blood**]. She's really sick, but all anyone
cared about was fucking politics. Even though she could be
dying, the administrators have to decide whether she can stay
in the hospital because she's not a U.S. citizen and not eligi-
ble for any kind of insurance. I was told that it would proba-

bly be okay if we said that she was here visiting relatives and got sick, rather than that she came here for medical treatment, which was the truth. I was told not to write anything in the chart until administration had cleared it and that I shouldn't exactly lie, but I should tell the truth in a certain way, you know, make it sound like she's really more acute than she is. I can't stand buffing the chart [**buff: polish to improve an appearance**]; it really bothers me. But fortunately, I had blown the cover the first day when I wrote in my admission note that she had come to the United States specifically for medical care. A lot of other people said the same thing in the chart, so it wasn't all my fault. Anyway, Loomis, the head of Adolescent Medicine, spoke to some big cheese and got the hospital to foot the bill. It was really nice of him, actually, and it was nice of the hospital, too. This's going to be a fucking thirty-thousand- or forty-thousand-dollar bill. But all this time, at least a half hour of attending rounds a day was being wasted on this bullshit.

Yesterday was just one of those bad days; I wasted the morning with the attending and the administrators trying to figure out what we were going to do with this kid. Then later on, I was told I was getting a patient with sickle-cell disease who was in painful crisis. Nobody bothered to mention to me that the patient was on the ward until the kid had been there for two and a half hours, lying in his bed down the hall and writhing in pain! I was so pissed I ran over to the head nurse and yelled, "Why didn't someone tell me this guy's been here so long? How come I haven't heard about this?" And she said, "Well, it's not our fault," and she looked at me as if it were my fault, that I should have somehow instinctively known the kid was there. I was so furious, it took me two hours to cool down because the main thing was, here's this poor kid, he's in agony, and he didn't have to be! What can I say? It was just one of those days.

I think I probably get flustered too easily. I shouldn't allow myself to get angry about these things. Okay, so the patient's lying there, writhing in pain. Complaining to the head nurse didn't do him any good, and it sure as hell didn't do me any good. I could have just very quietly, very calmly filled out an incident report that I wasn't informed of the patient's presence. I could have sent that down to administration and then, whoever's ass had to be cooked, let his or her ass be cooked! I shouldn't allow myself to get aggravated about stuff like that; there's more than enough other stuff to get aggravated about. That's easy to say, but I still get all fumy and angry whenever something like this happens because, really, deep down inside, I want to do a good job, and I don't want people to be suffering. It really pisses me off.

I also got into a fight with a lab technician last night. I admitted a teenager with leukemia at about 6:00 P.M. He came in because he had fever and the hematologists were sure he was septic [**had a bacterial infection in his blood; especially dangerous in patients with malignancies because their white blood cells, an important line of defense against invading organisms, are usually markedly deficient**]. He got sent up to the floor very fast; they had seen him in clinic but they hadn't even done any of the lab work. They were really worried about him, so they sent him up directly from clinic.

As soon as he got to the floor, I drew all the admission blood work out of his central line and sent it off to the lab stat. Of course, I included a CBC [**complete blood count**] with diff [**differential cell count: percentage of various types of white blood cells within the sample**], and I wrote the diagnosis on the lab slip so they couldn't blow off doing the diff [**the differential count requires some tedious microscope work; therefore it is done only in cases where there's an abnormal number of white cells or in cases of malignancy**]. When I called the lab about an hour later to get the results, they told me nothing was ready yet. I called back a half hour after that and they gave me the CBC, but the tech said they hadn't done the diff. I said, "What do you mean, you haven't done the diff? I checked off 'diff' on the lab slip and wrote the patient's diagnosis." He said, "We don't do diffs at night." And I said, "What do you mean, you don't do diffs at night? I need a diff on this patient; he's got leukemia, for God's sake, and he might be septic. I have to have a diff! This kid could die." And the guy said, "Lots of people die every day," and then he hung up. He fucking hung up on me!

I was ready to go down there and kill. But the senior I was on with said it wouldn't do anybody any good and that all I had to do was take another sample over to West Bronx's lab and they'd run it for me. I did that, and sure enough, the kid was neutropenic [**had a deficiency of the particular type of white blood cells most important in fighting off infections**] and septic. He's pretty sick. He may die. But the people in the lab don't care about stuff like that. They only know what the rules are.

The aggravations of being an intern are just endless. I would say nine out of ten interns say these same two words over and over again: "Internship sucks." I've heard that particular phrase so many times in the past week or two.

I mean, think about it: To the nurses and most of the rest of the staff, we're nothing but another piece of shit. To the nurses, anything that goes wrong is the intern's fault. Some-how they're always innocent and the intern always is wrong. And nobody around here seems to give a shit! There are really only three good nurses on this ward. The rest are worthless, lazy, uncaring shitheads who spend most of their time sitting around on their fat asses in the back room, watching TV and eating junk food. They don't want to do anything. They certainly don't want to take care of patients. They're so fucking, incredibly lazy! Anytime you ask them to do anything, even take a patient's temperature, they either take it as a racial slur or as a personal insult. Oh, my God, it's a federal case to get a temp done! I'm used to being in a hos-pital where the nurses were superefficient; they'd fall all over each other to get an order filled. I'm not used to this attitude.

And there's another problem: There's a major cultural difference. Here I am, this white, upper-middle-class Jewish kid, and most of the nurses are black, working-class women. We're from completely different worlds. God knows what they're thinking when they look at me and the rest of us, but I definitely get the feeling that they think we're a kind of annoyance they have to put up with. There's so few of them who really want to make the effort to work together. Oh, well, what can you do?

We got a really fascinating patient last night. She's this poor little thirteen-year-old Hispanic girl, very cute and ex-tremely suicidal. She was brought in by ambulance because she told someone she had taken a full bottle of asthma med-ication. She didn't have any symptoms and her theophylline level was zero, so she really hadn't taken anything. But she told the people in the emergency room that voices in her head were telling her to kill herself, so they admitted her.

Poor kid, she comes from the original scrambled family. She's under the care of her grandparents, each of whom has attempted suicide multiple times. She's with the grandparents because her mother is a drug abuser who severely beat the girl when she was younger. She lives in a complete fantasy world; she told me about it in vivid detail. But other than listen to her talk, there wasn't anything I could do for her. What she really needs is a psychiatrist. It's sad, it's really sad.

Karen's been here almost a week now. She's doing an elective in Manhattan. I've spent maybe six or seven hours with her, total. Next weekend she's dying to go to Philadelphia and she wants me to come along for this party a friend of hers is throwing. But that would mean I'd have to do every other night on call and I'd have to trade with Margaret, who would have to work an extra weekend day. I don't want to ask her to do that; it's not fair to her and it's also not fair to me. It would mean having to sleep on somebody's floor or something, and I'd come back and have to do the every-other. I'd be completely fatigued and I wouldn't even get to see Karen for five whole days. And that's just to try to go to a little party in Philadelphia. In other jobs, you'd expect to have every weekend off, and it wouldn't be such a big deal to go somewhere and have some fun. As an intern, you can forget it.

This past week Karen has had a very difficult time adjusting to my life. She's been really upset at my absence, at the fact that the first night she was here I fell asleep four times over dinner. She had just had this interview at a program for a psych residency and it hadn't gone well. The place seemed extremely disorganized, the people were disinterested, and they didn't know a thing about her. She was upset and she wanted to talk about it and all I could do was fall asleep; I'm worthless to her! So she was very frustrated. And since that night she's been getting angrier and angrier about the fact that two out of three nights I'm either away or asleep. Fortunately we have this weekend to be together. It's only Saturday afternoon now. I'm going to go to sleep for a while and then we'll have tonight and tomorrow together.

Friday, September 13, 1985

We are going down to Philadelphia tonight. I wound up having to go. I'm just waiting for Karen to come home. I don't know how I'm going to get through this, but I found myself in a position where I couldn't say no.

All the patients I'm taking care of now are psychopaths. Every adolescent in the Bronx is trying to commit suicide. They're either trying to do it by an overdose, by shooting themselves, or by starving themselves to death. The floor is chock full of anorexics and bulimics. There are two types: the "walkers" and the "liers." The "walkers" spend the entire day pacing up and down the halls. Whenever you need to find them, you just walk the corridors and there they are. They walk because they're trying to expend as many calories as possible, and this is about the only exercise they can get while they're in the hospital. They can't do their "jazzercise" four or five hours a day, so they just walk. The "liers" are worst off, though. They all look like concentration camp survivors; they're nothing but skin and bone. They're so debilitated, they can't do anything but lie in bed.

And there's nothing I can do to help them. I go and I try to talk to them, I try to reason with them about eating. They say they'll eat more but I know they're just doing it to get rid of me. They'll tell me anything they think I want to hear. Then they'll just go and do whatever they want.

I referred one of the psychopaths to my clinic today. A

patient of mine, one of my suicide-attempt kids. He's nuts, but he's really a good kid. I think he just needs someone to look after him. I can't do a very good job of that as an intern, but I can at least be a little bit more of a support system. While I was making the appointment for him, I was thinking, Do I really want to do this? Do I really need this much of a problem coming to my clinic every week for the rest of the year? I hope I don't regret it.

My med student is turning out to be great. She told me I was great, too. She said I really cared about people. That's nice; I'm glad she said that. We pat each other on the back, and that's important because nobody else verbally applauds us. At first I didn't want her to do any of my scut, that's not what med students are for, but she was always willing to help and eventually I just got used to her being there when I needed her. I always tried to teach her things while we were scutting out, kind of on a one-to-one basis: I taught her how to do a gram stain [**a test to identify bacteria in a sample of body fluid**] and then when I needed one done, she'd run to the lab and do it for me. I taught her how to read an EKG [**electrocardiogram**], how to put in a Foley [**a catheter passed through the urethra into the bladder, to monitor the output of urine**], how to put in IVs. She liked that. She's going to be a great doctor.

Sunday, September 15, 1985, 1:00 A.M.

I just got back from Philadelphia. I went down there with Karen and some friend. We went to a big party at Brad's house, but I couldn't stay for the bash; I had to come back home so I could be on call tomorrow . . . later this morning . . . Sunday.

Last night—Friday, actually—I got home postcall and

everybody came in after they'd gotten the car. They said they couldn't wake me up for ten or fifteen minutes; I wouldn't talk at all. I don't remember much, but I finally got out of bed and slept the whole way there. Soon's I got there, I went to sleep. I had a good time in Philadelphia; it was fun.

I talked to Ann while I was in Philly. She's a friend of mine from medical school. She's an intern doing pediatrics at St. Christopher's. Right now she says she just entered this phase of intense resentment and anger and depression and hatefulness. She hates her work, she hates her patients, and she especially hates anybody who's enjoying what they're doing. She says she doesn't think it's so bad because in three days she's going on vacation and she knows she can kind of get this way now. She says she didn't know what she'd do if she were in my shoes and had to wait another three months for vacation.

I'm starting to feel kind of apathetic, in a funny way. Does apathy follow depression? I don't have much more to talk about because I'm tired; tired and pissed. I'm also kind of sick of doing this diary, to be honest. I'm starting to feel kind of annoyed, I don't know why. I don't think I want to remember this fucking year.

Friday, September 20, 1985

I finally had a dream I remembered. I guess it's a good sign, but I think I'm better off when I forget them. The dream went like this: I was at work and the resident wanted me to check the potassium level on this patient who was taking a drug that depletes potassium. I kept making excuses, I don't know why, I kept putting off drawing the blood. Finally I was sitting in the library and a cardiac arrest was called. I went running in to find my patient dead and unresuscitatable. The rest of the house staff showed up and started yelling in unison that I had killed the patient. I woke up at that point in a cold sweat. I was terrified!

I can't exactly put my finger on what's wrong with me. I only have six patients now. Two of them are GORKs who don't make any extra work, and one is just a suicide attempt. But it doesn't matter. I still seem to get killed almost every night I'm on call. And then I come home postcall and fall asleep and Karen glares at me. I spend three hours with her every third night. Last Monday night, a friend from home came to visit. We went out for dinner. I fell asleep three times during the meal. When is it going to start getting a little better?

Amy

Monday, September 2, 1985, 11:00 P.M.

I just got Sarah to sleep and I finally have some time to myself. Today's Labor Day and I had the day off. We had a great time: We went to a zoo, we went swimming, and we just generally relaxed. It was something we all really needed! It was freezing last night; I had to put Sarah to sleep in a sweater. Summer's ending; I usually get depressed at this time of year, but this year is different. I'm actually happy summer's ending. It means I'll never have to be an intern in the summer again!

I started on 8 West on Wednesday and so far it's been a mixed experience. It's not nearly as calm as University Hospital. There are some interesting things going on but it's not so taxing that I can't get out early. So those things are good. But our resident this month is an idiot! He's completely useless. He can't teach, and when he tries, he gets it all wrong! On Thursday he told us that the peak age for SIDS [**sudden**

infant death syndrome: the unexplainable death of an infant] was six months to one and a half years. That's completely wrong! I know a lot about the subject, I've made myself crazy about it since Sarah was born, and I know the peak age is two to six months. I told Barry [**the senior resident**] what it really was and he stammered and argued for a while and finally said, "Let's just drop it, okay?" I looked it up in Rudolph [**a textbook of pediatrics**] after rounds and showed him I was right. He still hasn't admitted he was wrong.

He's also not very good technically. He couldn't start any of the IVs I've had trouble with and he can't get blood on kids after I've failed. So if he can't teach and he can't help, what good is he?

It's very upsetting having this guy in charge, but what can I do about it? So far what I've done is just ignore him. If I need help, I either ask Susannah, the other intern on our ward, or if she can't help, I go to the chief residents. The chiefs seem to understand our problem; they always come and help if we need it.

It's good to be working with Susannah. We were classmates at Schweitzer. She's got a ten-month-old daughter at home, so our motivation is pretty much the same: We both try to get out as early as possible. On days when I'm on call, I tell her to sign out to me as early as possible, and when she's on, I sign out to her as early as I can. Having this moron as our resident is mostly just a pain in the neck for Susannah and me. I feel bad for David, our subintern, and the medical students. They don't know how little they can trust Barry, and they might just believe some of the things he says. What this means is that Susannah and I are going to have to do some extra teaching this month to undo the damage caused by our resident.

I was on call Saturday, and it was quiet. I spent most of the day sitting in the residents' room watching TV. I got only one admission, in the afternoon, but we just about had to drag the patient up to the ward. It was a six-week-old who came to the ER with a history of a fever, a cough, and an eye discharge. The baby had been seen in some other ER the day before and was sent home, if you can believe that! Even I know that anyone under two months of age with a fever gets admitted to the hospital for IV antibiotics. But this bozo at the other hospital actually told the mother the baby was fine, he just had a little cold, and she should take him home. She brought him to Jonas Bronck the next day because he still had the fever. When the guys down in our ER heard it was a FIB [**fever in baby**] they pounced on him, did a complete workup, and got ready to admit him.

I wound up getting a few hours of sleep in the on-call room. The place is really disgusting; it's a tiny room at the end of 8 East with nothing in it except two sets of bunk beds and a telephone. Usually they don't provide pillows or blankets, so you have to steal linens from the clean-laundry cart, and you can do that only if the nurses like you or if they're not looking. They know me, so there wasn't any trouble. So life's not bad. At least so far.

Sunday, September 8, 1985

This is the end of another terrific weekend. We spent yesterday and today with my father in New Jersey. Because of the way the schedule worked out, I had Sunday and Labor Day off last week and this whole weekend off. I feel like a banker, not an intern. And I've actually had a chance to spend quality time with Sarah without the sitter being around.

Marie and I are getting along fairly well. We're not exactly best friends, but I think we at least understand each other. I guess I've come to accept some of her mishegoss **[Yiddish for craziness]**, and she's come to accept some of mine. I was concerned for a while that Sarah would come to think of Marie as her mother and me as someone else who happened to spend a lot of time around the house, but it's clear she knows who I am and who Marie is. Sarah's doing a lot of things now, rolling over front to back and back to front. She doesn't lie still for a second; we can't leave her alone on the bed anymore or she might roll off onto the floor. Then I'd have to take her to the ER for head trauma and they'd probably report me to the BCW.

I did bring Sarah to work with me last Wednesday. She had an appointment with Alan Cozza, and since his office is right off the ward, I figured we could get it in before work rounds. Alan told me Sarah's doing fine. She weighed over fourteen pounds. She's nearly doubled her birth weight in four months! And she got her second DPT **[diphtheria, per-tussis, and tetanus immunization]** and OPV **[oral polio vaccine]**. I had to leave the room when Alan stuck her. It definitely hurt me more than it hurt her. Now I can understand how the parents feel when we stick their kids. Sarah cried for about twenty minutes after he finished. And that's after pre-treatment with Tylenol!

Anyway, before rounds started, I sat in the residents' room with Sarah on my lap, and all the other interns and residents came over and oohed and aahed over her. A lot of them didn't even know I had a baby. It's so strange! Here are people I work with every day, I even sleep in the same room with them every third night, and they don't even know I have a baby! Of course, I don't know what's happening in their lives either. But this job tends to do that to you, it brings you into intimate contact with people who remain total strangers.

The work on the ward still is pretty easy. I've been getting out between three and four on the days I'm not on call; so has Susannah. The subintern's been staying until six or seven every night, but of course he's not really sure what he's doing yet, so naturally everything takes him longer. He also thinks he has to stay late to get something out of the rotation.

Everything would be perfect if it wasn't for that idiot Barry Bresnan! He really is dangerous. I hate work rounds in the morning. We do nothing but argue for an hour. He's got some very strange ideas about medicine, and most of the time they're wrong. One day last week, Susannah admitted a five-year-old with a hyphema [**bleeding into the anterior chamber of the eye; dangerous because it can lead to blindness**]. The boy had been hit in the eye with a baseball. Susannah did what the ophthalmology consultant told her to do: She put him in a private room, patched his eye, and kept him sedated so the eye would not get reinjured and bleed again. But on the rounds, Barry said there was no reason to sedate these kids; in fact, it was dangerous for some reason he never explained. He told us to stop medicating the boy and just let him run around and do what he wanted. Susannah told him he was crazy, that if the kid were to rebleed into his eye, it could cost him his vision, and that she was going to continue the medication because that was what ophthalmology wanted done. Barry yelled at her that ophthalmology wasn't running the ward, he was, and this is what he wanted to do. At that point Susannah figured it was useless to argue with him. She said "All right" and left it at that. She kept the patient sedated though.

Later in the day, when the chiefs made rounds with Barry, they told him the boy had to be kept sedated or else he might rebleed. So Barry came back to Susannah and said he'd changed his mind and she should start sedating the boy again. She told him she'd never changed the order in the first place, and that really pissed him off. But what could he do? He had been wrong and we had been right. He couldn't very well go to the chiefs and complain that the interns weren't following his orders, because his orders were wrong! He's so stubborn and so stupid. And he's dangerous. He could cause a lot of trouble for our patients. It's frightening!

Friday, September 13, 1985

I am really angry! That jerk is continuing to find ways to torment me. Every time Susannah or I make a decision, every time we try to do something to help one of our kids, he comes and tells us we're wrong and we have to change things, and we argue and get into a big fight. And it always turns out that we're right and he's wrong! It never fails! I don't know how he can be a resident and know so little! It's actually scary!

Today was the worst so far. We started off the morning fighting about a kid with asthma I had admitted last night. Nothing earthshaking or exotic, just a simple, straightforward eight-year-old asthmatic, something we see every day on the wards, and he found a way to screw up the kid's care!

This was a kid who gets admitted to the hospital four or five times a year. They gave him a minibolus of aminophylline in the ER. [**Aminophylline, a drug that dilates the breathing tubes, was, at the time of *The Intern Blues*, the main-**

stay of asthma therapy. In severe asthma attacks it was given by vein either in boluses, when a large dose is given once every six hours, or by constant infusion or drip; more recently, newer, inhaled medications have supplanted aminophylline in the treatment of asthma]. I started him on a one-per-kilo drip [a drip containing one milligram of aminophylline per kilogram of body weight per hour, the dose needed to maintain the blood aminophylline level] after checking the old chart and finding that that's what it took to maintain his level. I drew levels [blood samples to determine the amount of aminophylline in the blood] after the bolus and four hours after starting the drip, and they showed he was in the therapeutic range. By this morning he was much better, but still he was wheezing a little. On work rounds Barry asked me what we had done. I told him and he said, "No, you did it all wrong, that's not the way to figure out what dose of aminophylline to give an asthmatic." He then recalculated everything using this strange formula I'd never seen before and told me that we should have started him on a 1.5-per-kilo drip. I told him he was crazy, that if we put the kid on a 1.5 drip he'd get toxic [develop blood levels of aminophylline in the toxic range; signs of aminophylline toxicity include nausea and vomiting and convulsions] in a matter of hours and I certainly wasn't going to do that to my patient. He told me I was wrong, that he'd always used this formula and he'd never had any trouble with toxicity. That's when Susannah told him he probably just had been lucky in the past because she was positive that if we changed the dose to 1.5 per kilo, the kid would be vomiting by noon. She also happened to mention to Barry that she thought he was both full of shit and dangerous and that we'd all be better off without him. Although she and I had been thinking all of this since the very first day of the month, neither of us had said it to him before. He yelled back at us that we were the ones who were full of shit and that we could say whatever we wanted about him, but he was sure he had never seen two interns who cared less about their patients. I got really angry at that point and asked how he could say that. He said something to the effect that he had never heard of interns who left the hospital at three o'clock every afternoon.

Well, we were yelling loud enough at that point to get the chief residents out of the ICU, where they were rounding. Susannah said something about the fact that she and I were able to get out early because we had learned to be very organized and that maybe Barry's problem was that when he'd been an intern he hadn't learned anything and that was why he'd turned into such a bad resident.

That was when the chiefs separated us. Jon Golden pulled Susannah and me down the hall, and Eric Weinstein took Barry into the residents' room. Jon asked us what had happened, and I related the incident from start to finish. Jon said that Barry, as usual, was definitely wrong and that we were definitely right. He told us they were having a lot of problems with Barry, that he really did have some problem with his knowledge base and that, on top of it, he had no confidence in himself. "Of course he has no confidence in himself!" I told Jon. "He doesn't know anything! There's no reason for him to have any confidence!" Jon told us that may be true, but they were stuck with him, and we would have to try to make the best of the situation. He also told us we should try to be nice to him because it might help him with this confidence problem. That's not exactly what I wanted to hear and it wasn't what Susannah wanted to hear, either. She said there was something really wrong with all of this, that this guy was dangerous and he shouldn't be in a position where he had the chance to harm the patients. I said I thought he should be thrown out of the program. Jon kind of groaned and said that he really wasn't that bad and that we just should give him a chance. We just walked away at that point. There was nothing else we could say. We weren't getting through.

I don't know what Eric told Barry, but he came out about a half hour later and didn't say a word to us. We finished work rounds without him and made all the medical decisions ourselves.

At about three o'clock I was signing out to Susannah in the nurses' station. I didn't have much going on; it's been kind of quiet. I only have six patients, and two of them are chronic AIDS patients. [**The back rooms on 8 West are reserved for children with AIDS and AIDS-related illnesses. On any given day there are four to six patients residing there. Most are not sick; they have come to live in the hospital because there's no other place for them**]. While I was signing out, Barry came up to me and said, "What are you doing?" I told him I was signing out and he said, "Look, you can't sign out at three o'clock." He said there had to be some work I still had to do. He said that I had a responsibility to teach the students and obviously they weren't getting taught anything if we were all going home at three o'clock.

I didn't answer him; I just continued signing out. After a while he said, "Didn't you hear me? Why don't you answer me?" I told him I didn't answer him because I didn't have anything to say to him. He looked really hurt and walked away. A couple of minutes later, Jon Golden came into the nurses' station and asked if he could talk to me in private. I had finished signing out and was getting ready to leave. He took me into the chief resident's office and said, "I know things aren't going well on the ward, but don't you think you could try to be nice to him?" And I said, "Why? He's not being nice to me!" I told him about what Barry had just said to me and he kind of sighed and just asked if, in the future, I would just play along with Barry, at least make him think I was being friendly. I said I'd try. And then I got up to leave and on my way out, I ran into Barry again. I said good-bye to him and this time he ignored me! Unbelievable!

There are two more weeks to go on this ward. If it hadn't been for this jerk, this would've been a very nice month. As it is, I can't wait until it's over!

Monday, September 23, 1985

Thank God the month is almost through. I haven't recorded anything in over a week. It's been too depressing. I hate going to work every morning and spending the whole day fighting.

It wasn't as bad in the beginning of the month because we weren't all that busy. But last week the place started to fill up and there are a lot of complicated patients around who need a doctor who knows how to make decisions. Susannah and I don't know enough, and since we've stopped talking to our resident, there's nobody to turn to except the chief residents, who aren't all that thrilled about being bothered with our trivial stuff every five minutes. But these cases are complicated and we can't manage them alone!

I now have nine patients. They include three asthmatics, one of whom was really sick and almost needed to be intubated [**had a breathing tube placed through the larynx and into the trachea to facilitate artificial ventilation**]; a four-year-old girl with nephrotic syndrome [**a condition in** which the kidney fails to retain protein; the protein spills out in the urine and the patient becomes protein-deficient, which leads to severe swelling of the entire body]; an eight-year-old girl who's GORKed out after being in a fire and inhaling a lot of smoke [brain damage occurs in patients with smoke inhalation usually because of carbon monoxide poisoning] and whose only sign of brain activity is her daily convulsions; two FIBs on antibiotics, one of whom probably has meningitis; and Winston and Salem, my AIDS twins (it's still hard to believe anyone would name a pair of twins "Winston" and "Salem," but there they are, on my ward). And even they're not doing so well. Susannah's got ten patients and the subintern's got six and we have to cope with an idiotic resident.

What happened today with Winston and Salem is a perfect example of what's going on. Salem developed some pimples on his chest this morning. I was pretty sure it was chicken pox and I had Susannah look at them and she confirmed it. And Winston's had a cold for the past day or two, now he's probably coming down with it also. Susannah and I got very worried. We both know that varicella [the virus that causes chicken pox] could kill them [varicella, usually the cause of mild illness, can cause an overwhelming infection in persons who are immunodeficient], but we didn't know what to do. Neither of us wanted to ask Barry; we knew he wasn't going to have anything helpful to say. So I went to Jon Golden right away and he said it probably won't be a problem because both of them are getting gamma globulin treatments [a treatment modality that has had some success in children with AIDS] and that the gamma globulin had antibodies to varicella so they probably wouldn't get an overwhelming case. Going over Barry's head worked out okay this time, but what's going to happen on those nights when there's nobody else around except him?

At least I have Sarah and Larry to come home to, and that makes me feel much better. Even after a day like today, ten minutes after I saw my baby, I was back to my old self. You know, when I started this internship I was concerned because I didn't know if I'd be able to be a good mother and a good intern. Now I don't understand how you can be a good intern without also being a good mother. I'll tell you, though, if I had to choose at this point whether I wanted to be a mother or a doctor, it wouldn't be a hard decision to make.

Sunday, September 29, 1985

I don't have much time. We're waiting for some friends to come for dinner. What can I say? I made it through the month, but it took a lot out of me. I'll never speak to Barry Bresnan again, that's for sure, but I survived it. It's now three months down, nine months to go. Eight if you count vacations. I'll survive it, I'm sure of it. I've managed to make things at work relatively easy for myself, but trying to be a mother and a doctor is taking a lot out of me. So far I think I'm doing a reasonable job at both; I just hope I can keep it up.

Mark

Sunday, September 1, 1985

Ah, one day on call in the emergency room and suddenly I remembered exactly what it was that made me become a doctor in the first place. Yes, I'm sure the reason I became a doctor was so hundreds of mothers and fathers who don't speak a word of English could curse at me in their native tongues while expecting me to cure their little darlings completely within minutes. What a rewarding experience yesterday was!

Actually, it wasn't all that bad. It wasn't too busy. I spent most of the afternoon taking care of a six-week-old who came in with fever and a cough. Everyone, even a lowly intern, knows that a six-week-old with fever is an automatic admission. But apparently that's not something that's taught to the interns at BEPI [**internese for Bronx Episcopal Medical Center, a voluntary hospital in the South Bronx**] because this kid had been seen there a day before, had been started on amox [**amoxicillin, an antibiotic**], and sent home. Sent home, for God's sake! The mother brought him to Jonas Bronck only because his fever hadn't gone away and she happened to be visiting a friend in Jonas Bronck's neighborhood, so she decided to stop in and spend a few hours watching TV in our waiting room for a change. So not only did I have to do a whole sepsis workup [**spinal tap, blood, and urine for cultures**] on the kid, I also had to explain to the mother that the baby would have to be admitted. This must have sounded a little strange to her. After all, the kid was better than he had been the day before, and if he was so sick, why didn't the doctor at BEPI say that he needed to be admitted?

What can you say in a situation like that? "Oh, the doctor at BEPI didn't admit him because he was a malpractice major at Our Lady of the Offshore University of Medicine and Hair Design"? It's hard to tell a woman that the last doctor to whom she trusted her kid was a moron who might have killed him! So I told her that some doctors are more conservative than others about these things and that keeping the child was mainly a safeguard. That's the first time I ever used the word "conservative" as a euphemism for "competent."

Anyway, the story doesn't end there. That was really just the beginning. I finally convinced her that the kid needed everything, including a workup and admission. I went ahead and drew the blood and did the spinal tap. I also did my first bladder tap, which is a pretty nasty procedure [**to do a bladder tap, a needle attached to a syringe is passed through the abdominal wall and plunged downward until urine appears in the syringe**]. So I finished all that and I told the mother to go back out to the waiting room and hang out until someone from the ward came down to pick her up. Up until that point, everything was fine.

I called the intern upstairs and it took her about a half hour finally to get down to the ER. It was Amy Horowitz. I sent her out to the waiting room to find the kid but she came back a few minutes later and said, "They're gone. Do I still get credit for the admission?" [**Admissions are distributed evenly to the two interns taking call on the inpatient wards at Jonas Bronck; they alternate, admitting every other one.**] I went out to look and I couldn't find them either. It looked like they had vanished into thin air. Bob Marion was the attending and he told me to try to call the woman at home using the number listed on the ER sheet. So I called and I got some woman who claimed not to be able to speak any English. I asked for the kid's mother by name and this woman on the phone said something about "no speeka de English." So I got one of the nurses to translate and she told me that the woman was saying that she had never heard of anybody by the name that was listed on the sheet. I figured I dialed the wrong number, so I hung up and tried again. I got the same woman on the line and we went through the same routine again. I was sure I was talking to the mother of the patient; I was positive she had decided that she didn't want the kid admitted, that she had never believed me in the first place, and that she had gone home and would deny that she had ever been to Jonas Bronck. It made me crazy! I told Bob about this and he got angry, too. He called the number himself and told the woman in Spanish that the baby could be very sick and if she was lying about it and the kid did die, it wouldn't be his responsibility, it would be on the mother's head and not his. Amazing! I had never heard anybody using the concept of Jewish guilt translated into Spanish. It didn't work, so it may have lost something in the translation, because the woman told him he was crazy, that she had never heard of the people he was talking about, and if any of us tried to call her back again, she'd call the cops and have us all arrested.

Bob hung up, and his face was bright red. He yelled that I'd better call the police and have them go out to the address listed on the ER sheet and bring the baby back, dead or alive. That is, if the address listed was correct. So then I started trying to contact New York's finest. Jesus, what a day!

I called the precinct house in the neighborhood they lived in and spoke to the desk sergeant. He took all the information but didn't know what to do with it so he tried to connect me with his supervisor but somehow the phones got disconnected. So I called back and got somebody else, who reconnected me with the desk sergeant who again tried to connect me with the supervisor. While I was on hold, the senior resident came up to me and said, "The charts are starting to back up. When are you going to be done with this nonsense?" I started to answer but then someone picked up and I never finished my sentence. I had to give whoever was on the phone the information all over again, and he said, "Well, you understand we can't just go out there and arrest them. We can't drag them in against their will. If the woman doesn't want to come, we can't force her." I told him just to do his best; he said he'd try and asked for their address. When I gave it to him, he said, "That's in the projects, isn't it?" I didn't know, so he looked it up and said it was. Since it was the projects, it was out of the city cops' jurisdiction. He told me I'd have to call the Housing Police.

So I'm sitting in the emergency room, the patients are starting to pile up in the waiting room like the planes over Kennedy International Airport on a Sunday afternoon, and I'm getting a civics lesson in the structure of the New York

City Police Department bureaucracy. I was pretty pissed off about all this. I called the number of the Housing Police and went through the same business again but finally a sergeant told me he'd send out a squad car to see what they could do.

I figured that phase of this mess was over and I went to pick up a new chart from the triage box but before I could even make my way to the microphone to call the next patient in from the waiting room, one of the nurses came running up to me and told me to follow her. She led me into the adult ER and there, in the holding area, lying on a stretcher, was the woman with the baby in her lap, both dead asleep. She had lost her way out to the waiting room, had come across this nice, cozy, empty stretcher, and had decided to use it to catch up on some sleep. I couldn't believe it! I went back to the ER, called Amy, and told her to come down and pick up her patient.

Then I remembered that a crack unit of the Housing Police was preparing to swoop down and make a raid on the woman's apartment. I figured it was my civic responsibility to try to prevent them from going on this wild-goose chase, but when I called, I found out I was too late; a squad car had already arrived at the projects and they couldn't be called back. About ten minutes later, two huge Housing Policemen marched into the ER leading this panicked little eight-year-old kid. "He said his mother wasn't home," one of the cops said. "He said she was still here with the baby." Bob and I explained the story to them very nicely. They weren't even upset. They didn't threaten to slug us in the mouth or anything. They even volunteered to take the boy back home. So the story had a happy ending. What a strange job this is!

I waste a lot of my time in the ER talking on the phone. If it's not the police, it's the labs. The lab techs don't give a shit about anything. There's this rule that before they give out any information to a house officer, they have to torture him verbally for a while. And if it's not the labs, it's a consultant from one of the subspecialty services who wants to avoid coming in to see a patient at all costs. And if it's not a consultant, it's the Bureau of Child Welfare. BCW's the worst! I never really knew what the term "terminal hold" meant until I made my first BCW referral. If putting people on hold were an Olympic sport, the BCW would be the gold medal winner. They can keep you waiting for twenty or thirty minutes without breaking a sweat. If all of us house officers could just see patients and not deal with the rest of the bullshit, there'd be no waiting to be seen in the emergency room.

Thursday, September 5, 1985

A week in the Outpatient Department has done me a lot of good. I'm more relaxed now, less on edge than I was on Infants'. Take Hanson, for instance. When I think about him I actually find myself laughing at some of the things that happened two and three weeks ago, how angry I got when he crumped. I guess with the passage of time, there'll come a point where I actually have fond memories of him. I can imagine: "Ahh, that Hanson, what a wonderful child, what I learned from him, how I wish I could take care of him again!" This is all kind of frightening. I think it's these kind of warped remembrances of internship by people in charge of training programs that keep us working every third night!

Sunday, September 8, 1985

I had this whole weekend off, and Carole and I sat down and actually tried to plan out my vacation. Carole can get only the second week of it off, so I'll have to figure out something to do the first week. We argued about it for a while; I

wanted to go to some quaint New England village and just
sack out, but Carole didn't like the idea of wasting a whole
day in the car getting there and another whole day coming
back. She wanted to go to some hotel in the Poconos. So we
compromised: We're going to a hotel in the Poconos. Ahh,
what the hell, it doesn't matter that much to me. Anywhere
we go is fine. As long as there's a place to sleep and it's far
away from here.

Working in the ER is fine. I'm seeing a lot of patients,
nothing major or earth-shattering, just routine walk-in clinic
pediatrics. I think I'm holding my own. I'm learning to do a
lot of things I never knew how to do before. I'm even learning a little Spanish. I now can have three-year-olds refuse to
open their mouths, stick out their tongues, and say "Ahhh"
in two languages. Who says internship is not an educational
experience?

Hanson keeps repeating on me like a bad hot dog you get
from one of those umbrella carts. I got a call from Jennifer
Urzo, the intern who picked him up when he came back to
Infants' after being discharged from the ICU. She said he was
doing fine, hadn't crumped in over a week, and wasn't he the
most adorable thing I'd ever seen? I restrained myself. She
said they're starting to think about discharging him and
wanted to know if I had any idea how to get in touch with his
mother. I told her I'd never set eyes on the woman, that she
never once showed up during August, but if she was able to
contact her, she should give me a call because there were a
few things I'd like to say to her!

I have nothing much else to say. I can't wait till Thursday.
Imagine, I'll be able to stay in bed past seven o'clock two
days in a row!

Bob

People who live outside of New York City carry a vivid image of what the Bronx is like. That image is based on a picture that appeared in newspapers around the country, a photograph of then-President Jimmy Carter standing amid the burned-out rubble of Charlotte Street. But the poverty and the dilapidation of the South Bronx is really only one view of the borough. At the opposite end of the spectrum, at the northernmost part of the Bronx, there's Riverdale, one of the wealthiest sections of New York City. And in between the North and the South there are numerous middle-class neighborhoods, each with its own special character and ethnic flavor. The people from all these communities share one thing in common: They receive their medical care at the hospitals in which our interns and residents work.

But the poor children of the South Bronx are by far the Pediatrics Department's biggest customers. They're the ones who crowd the waiting areas of our emergency rooms and fill past capacity the beds of our hospitals. Many of the problems from which these children suffer are directly related to the extreme poverty in which they live: They get anemia and growth failure because of inadequate diet, lead poisoning because they eat the paint chips that fall from the ancient walls and ceilings of their apartments, and asthma from inhaling the polluted air that hangs over the South Bronx like a deadly cloud. Other medical problems are indirectly related to the poverty: As young children they're physically and sexually abused and abandoned by their angry and frustrated parents and other adult caregivers; as adolescents, unable to find jobs or stimulation, seeing little to look forward to, they turn to drugs and sex, having babies when they themselves are still children and getting infected with venereal diseases and the human immunodeficiency virus in the bargain. And all these problems fall into the laps of our house officers, who have to work doubly hard to figure out how to relate to these abused children, sixteen-year-old mothers, and adolescent crackheads while very often functioning simultaneously as doctors, psychologists, and social workers.

The vast majority of the patients who come to the emergency rooms at Jonas Bronck and West Bronx hospitals are black or Hispanic. The vast majority of the house officers are white. Sometimes it's difficult for the patients to relate to these doctors who know very little about how poor people live or what makes them tick. When I was a resident, I saw a six-year-old boy who had come to the emergency room with a fever. I diagnosed an ear infection and prescribed an antibiotic in liquid form. I carefully instructed the boy's parents to give one teaspoonful of the medicine every six hours around the clock and to keep the container refrigerated. A week later, the child was back with the same symptoms. When I asked if they had given the medication as prescribed, the mother explained that she had tried, but since they didn't have electricity in their apartment, they couldn't keep the stuff refrigerated, nor was there enough light at night to measure it out. The therapy was bound to fail because I had no understanding of the social situation. It became necessary to treat the child with a capsule form of antibiotic to treat the infection effectively.

Dealing with these kinds of social situations is a huge problem for our house officers. As I've already mentioned, frequently social conditions are the direct cause of our patients' illnesses. The house officer can treat the asthma or the lead poisoning, but after the child is better, he or she will be sent back home and most likely will encounter the same environmental hazards that caused the illness in the first place. To provide really effective care, the home conditions would have to be altered, a monumental and frustrating task. Our overworked house officers wind up having to settle for treating the symptoms rather than the underlying disease, an unsatisfactory but necessary compromise.

Another big problem our interns and residents face is caused by their lack of understanding of their patients' cultural background. For example, people from certain areas of Puerto Rico have a very complex belief system based on hot and cold. Some illnesses are considered "hot," some are thought to be "cold." Similarly, remedies are believed to be effective for either hot or cold illnesses, but usually not for both. If a doctor prescribes what turns out to be a "cold" remedy for a "hot" illness, not only will the parent of the patient not use the medication, but also he or she will lose all confidence in that doctor.

Many of our patients speak no English and must rely on a doctor, nurse, or other patient to translate for them. In the emergency room, this slows down the doctors' progress through the pile of charts of patients waiting to be seen, lengthening the waiting time dramatically. The net effect of all these problems is that hostility builds between patient and caregiver.

Sometimes there's a great deal of hostility. Many of our patients use the emergency room as a kind of walk-in clinic, showing up at all hours of the day or night for problems the house officers consider trivial: belly aches that have been going on for three or four months, headaches for which no aspirin or Tylenol has been tried, mild gastroenteritis, complaints that the interns and residents know could be handled in a clinic setting, over the phone, or by the parents just using a little common sense. Since no one who shows up at the door of the ER can be turned away, the house staff winds up having to see these patients, getting backed up, and ultimately losing sleep because of what they consider this abuse of the system. And when one is chronically sleep-deprived, this can easily turn into resentment and anger, the ultimate effect of which is that the house officer will come to view the patient as an enemy.

Although hostility might exist between patient and doctor, it's nothing compared with the hostility that exists between a doctor and some of the other members of the staff. The best example of this is the relationship between the interns and the people who work in the laboratories. The intern knows, almost instinctively, that fighting with the lab technicians will only bring him or her misery. No matter what happens, arguing with a technician is a fight the intern can only lose. The lab technicians, after all, hold the key to the completion of the scut list; without the results of lab tests, the intern can never go home. But sometimes it's impossible to hold back.

As a house officer, I managed to hold back every time but one. When I was a senior resident, I was taking care of a sick preemie who was scheduled to go to the operating room the following day. It was my job to make sure that the child was pre-op'd, which included sending a specimen of blood to the blood bank for typing. The intern on call that night tried six times to get blood from this poor baby, who seemed to possess no visible veins in his entire body. I tried and failed four more times. Finally, on my fifth attempt, I succeeded; I managed to get about two cc's of blood.

I put the precious specimen in the proper tube, gave it to the medical student who had volunteered to carry it down to the lab, and went on with the rest of my work. About two hours later, a nurse happened to mention to me that the technician in the blood bank had called and said that because the tube and lab slip had been signed by a medical student instead of a doctor, it had not been acceptable. He had tossed the specimen in the trash and now was demanding a second sample.

Needless to say, I got angry. Trying to hold myself back, I ran down to the lab. I explained to the technician how difficult it had been to get that blood; I described in vivid detail how small and sick the baby was. He told me he was sorry, but rules were rules; unless a person with M.D. after his or her name has signed in exactly the right places, he had been instructed to dump the tube.

I got angrier. I started searching the trash cans in and around the blood bank. The guy caught me and said, "It won't do you any good. I poured it down the sink."

That's when I really lost it. Three years' worth of repressed anger at laboratory personnel was immediately released in a single, spectacular tirade. I cursed out this technician, I cursed out his mother, his father, the rest of his family and friends; I went on for at least ten minutes. He didn't say a word, he just continued doing what he had been doing when I had first appeared. When I finally ran out of steam, I went back up to the ward and tried to get another sample of blood from the baby. It took only three attempts this time, but I got it, carefully signed the tube and lab slip, and dropped it off myself.

This kind of explosion is not uncommon. Lab technicians have a great deal of work to do and have a lot of people on their backs trying to get results. It's impossible to make everyone happy, so frequently no one is made happy. Many house officers will tell you that dealing with lab technicians is the most aggravating of all their jobs.

Perhaps all of these little aggravations make working as an intern in the Bronx more difficult than working in some other area of the world. Battling the environment, the patients, and certain members of the hospital's ancillary staff while chronically overtired is no mean feat. Year after year our house staff does it, and they learn to do it well. But the question still remains: Is it all worth it?

Andy

Monday, September 30, 1985

I guess I'm starting to get sick of talking about internship. I don't talk about it to other people very much anymore; I used to think maybe I shouldn't talk about it, maybe people wouldn't be able to understand what I was saying, but now I just don't want to anymore. Nobody fucking understands.

I'm over at University Hospital now, into my fourth month. I'm going to lose my mind before I make it to my first vacation. I have three more months to go, so I'm now only slightly more than halfway there. I've finished a quarter of my internship year. If you include vacations, I've finished a third of it. Yeah, only eighty more nights on call, right?

My apartment has become disgusting. I have so many roaches, they're crawling all over the place. I turn on the faucet and the water goes in the sink and the roaches come cruising out. They all dive off the edge of the sink right onto you if you don't move fast enough, because they're trying to get away from the water. Too many roaches, no one to talk to about roaches; no one to talk to about roachy thoughts. I wish there were people on my team I felt buddy-buddy with, like last month. But there aren't any.

Thursday, October 3, 1985

I'm really tired. I don't have anything to say anymore. I've been feeling depressed and apathetic and ground down lately. Tomorrow I canceled plans to go out with a friend. You know what I want to do instead? I want to be by myself. Isn't that weird? Tomorrow's a precall night, the only time I ever feel even a little rested, my only chance to have a little fun. And all I want to do is be by myself.

There were a couple of times today when I thought I should never have become a doctor. I just don't have what it takes. I don't know, it must not be true, people say I'm really good sometimes. But that's how I was feeling.

Putting in too many IVs . . . yeah . . . yeah.

Friday, October 4, 1985, 1:00 A.M.

Should I tell you about the baby who died the other night when I was on call? Should I tell you? Another one died, this one right in front of my eyes. A DNR baby [**DNR: "do not resuscitate": DNR orders are written only after careful consultation with all parties involved in the patient's care, including the child's parents**], very sick. She was born with multiple congenital anomalies and they couldn't figure out what caused them. She had terrible heart disease and it was only a matter of time.

She had been admitted the day before I was on call

because she was severely hyponatremic [**had a marked deficiency of sodium, an essential electrolyte, in her blood**].
As soon as she hit the floor, all these consulting services came
to see her: genetics, renal, neurology, and endocrinology, you
name it, they came by. Harrison, the intern who had admitted her, signed her out to me, saying that nobody knew what
the hell was going on with her but that it didn't really matter
because she was DNR, and if she crumped, I wasn't supposed to do anything but stand by and watch. The only scut
he told me to do was that if she died, I was supposed to call
the neurologist so they could do a brain biopsy [**take a specimen of brain in hopes that studying it would suggest a
diagnosis**].

It hadn't been a bad night; things were pretty quiet.
Then at about eight o'clock, a nurse came and told me and
the resident that the baby's breathing had stopped for a few
seconds but then started again. The resident and I went in to
look, and sure enough, she was having these long pauses in
breathing. I did a quick physical exam and couldn't find anything specific that was different except the breathing pattern.
We asked the nurse just to watch her, and we went back to
the nurses' station.

About five minutes later, the nurse came running out
again, saying, "Now she's not breathing at all!" We went
back into the room and found that the nurse was right; the
baby wasn't breathing; she was also bradycardic [**had a slow
heart rate**]. The resident grabbed an ambubag and started
bagging her and the heart rate started coming up again. Then
all of a sudden it hit me: We were resuscitating a DNR baby.
By that point it was too late: The heart rate had come back to
normal.

I felt bad; I mean, this might have been the baby's one

chance to die, and by resuscitating her, we kind of screwed that up. No telling how much longer the baby was going to hang on now. I went back in to talk to the father, who had been sitting in the room this whole time, to tell him we weren't going to do any more resuscitation. The father agreed.

About an hour later, the nurse came to tell us the baby wasn't breathing. We went back into the room, but this time we just quietly walked over to the bedside and listened to her chest. She still had a heart rate, but there wasn't much respiratory effort left in her. She was white as a sheet; I've never seen a baby that white before.

We stood over her like that for a while, occasionally listening to her heart, and finally, after about fifteen minutes, it stopped. The baby was dead. I looked up to the resident, expecting him to say something, and he just stood there with a goofy look on his face. I was thinking, You're in charge, you've got to say something. But he didn't say a word. It was very uncomfortable for a while, and finally I had to say it. I had to tell the father that the baby had died. There I am again, having to tell a parent that his kid was dead. I still have no formal training in it, but once again, the job fell to me. Why am I always the one? At least this time, everyone was expecting it, so it didn't come as a shock. But it still made my skin crawl.

I took the father outside the room and let him sit by himself for a while. Then I had the nurses come back and together we cleaned the baby up, took out all the tubes and stuff that had been in her. We swaddled her in a blanket and cleaned up the room. I had learned to do this in the NICU; after babies had died in there, the attending always tried to put everything in order before letting the parents spend time with their child. It made a lot of sense to me. So when we were all ready, the father came back in, I handed him the baby and sat him down on a chair. He held the baby, and we all left the room so he'd have some privacy. Standing outside, I could hear him cry. I started crying a little myself.

After a few minutes, I went to call the neurology attending. She was a total bitch; she yelled at me for not calling when the baby had stopped breathing the first time. I told her I had been told to call her after the baby had died, not when the baby was dying. I thought I was doing them all a favor, and all I got for it was a bunch of abuse. She told me I might have ruined any chance of making a diagnosis because of the delay. Shit!

Well, that was my last night on call. I've been feeling a little better, though. I wasn't completely depressed at work today. I actually enjoyed myself a little bit. I realized there are two things that make me enjoy work: sleep and not being on call. Being on call is the worst because when you're on call, even if you get very tired and you have tons of work to do, you still have to do it, there's no one around to help you out. It's very stressful. Being postcall is next worst because you're really tired and you always have a fair amount of work still to do, but you feel some relief because you're finally off the hook. Of course, the best time is when you're not on call and you're not tired, like today. I really liked that, it was really nice. Internship could almost be good if there were more times like today.

Sunday, October 6, 1985

I just got home. I'm postcall, I got no sleep, it was a busy night. I admitted four patients, which isn't bad, but there was an AIDS patient who was exsanguinating [**hemorrhaging;**

losing all his blood] on the floor, a renal transplant kid who was in the midst of an acute rejection crisis, and some other patients who were basically causing trouble. My admissions were hideously staggered throughout the day and night, and I didn't know what the fuck was happening.

I actually got a really interesting case last night: a little kid with argininosuccinicacidemia **[an extremely rare inborn error of metabolism caused by a deficiency of an essential enzyme that results in liver disease, neurologic dysfunction, frequent infections, and, often, death in early childhood]**. The kid is really sick, but he looks just like the Michelin Man. He's got layers of fat around his belly and arms and legs, and it looks like tires. It's hard to feel sympathy for him because I laugh every time I look at him.

Harvey Abelson, the director of pediatric home care, the service that manages all the chronically sick patients, is the Michelin Man's doctor. He's really nice and he's smart and he's a good teacher from what I can tell. Very intense, totally intense guy.

The transplant kid sounds like the beginning of a sad story. He got this new kidney in August. His mother donated it to him. There were some problems around the time of the transplant, but he pulled through. Now he comes in with rejection crisis. His blood pressure's about 180/140 **[normal for this patient would have been about 100/60]**. They're talking about having to take out the good kidney to bring down the blood pressure. The mother's beside herself.

Monday, October 7, 1985, 11:00 P.M.

I think I want to be Harvey Abelson when I grow up. He's hyper but he's a pretty impressive guy. He's right on top of everything all the time. You should see him working out what needs to be done with the Michelin Man. He knows everything! Yes, Harvey Abelson, that's who I want to be.

My brother got married today. He didn't want anybody else to be there, so he invited only a couple of people. My parents weren't invited, I wasn't invited, we were "uninvited." I wished I could have been there, but I respect their wishes. My parents, however, are major-league pissed off. My brother and his wife, they were going to do it in December but then at the last minute they decided to do it today. I heard about it, of course, from Karen. I don't know . . . there has to be a better way.

I had to stay late tonight to get my work done. My transplant patient seems to be better; his renal function's coming back and his mother's calmed down a little, too. But there was a lot of scut work to do so I had to stay until nearly eight. Now I'm going to bed. A solitary life, that's what I'm leading.

Let me check if there're any roaches in my bed. I've had roaches in my bed recently, roaches crawling in my bed. I woke up the other day and there was a dead roach underneath me. I must have rolled over and squished it.

I don't know how I'm going to make it until Christmas, when I have my vacation. I just don't know how I'm going to make it. It's just too long.

Some kid asked me what I was doing, who I was today. He asked me if I was a doctor and what kind of doctor I was. I told him I was a hospital doctor. Just a hospital doctor. Someday I'll look back on all this and cry. Has it always been this hard?

Tuesday, October 8, 1985

I've got to do something about this. I've been wallowing in this low- to moderate-grade depression for about the past month now, and that's all I'm doing is wallowing. I can't stand this feeling anymore. I've got to do something about changing my attitude. It's the only way, because if you can't go over and you can't bow out, you gotta go through it. And that's what I've got to do.

Thursday, October 10, 1985

I spent a long time teaching my medical student today, telling him about fluids and electrolytes. Very exciting! I feel like I don't know anything. I guess that's not exactly true, though. I do know how to teach about fluids and electrolytes. But that's about it.

And I got myself reorganized again. Started using my daily flow sheets again on all my patients, something I hadn't done for the past week because I was too depressed. I didn't give a shit, and I never really knew what was going on, and that made me more depressed and so I gave less of a shit. Getting organized should help. Getting organized will make things better.

I'm feeling kind of horny these days. Doesn't help having Karen two hundred and fifty miles away. Kind of get to eyeing the nurses, you know? But I don't think I'm going to follow up on any of that. Nope! Think I'll just stay true blue and all that stuff. It's just a couple more weeks; I think I can make it.

It turns out that Harrison Boyd, the other intern on our team this month, is completely insane. He's got a very funny, terrible sense of humor. Very bad jokes, the worst! And Laura Santon, who is always happy, actually looked depressed today for the first time. Maybe she was just spacing out, but she looked kind of upset. I was surprised.

Had pizza for the hundred and fifty thousandth time for dinner tonight. Seems that's all I ever eat around here, pizza.

Missed the shuttle [**the bus that travels between Mount Scopus, Jonas Bronck, and University Hospital**], had to take a cab home, and got a free, unguided tour of the Bronx, because they always take some strange route. Very interesting, the Bronx at night. Very exciting. I could have done without it.

Sunday, October 13, 1985, Morning

I've had so many nights of sleep in a row, I practically don't know what to do with myself. Yesirree, I was on call Friday night and I got seven and a half hours of uninterrupted sleep, breaking all records heretofore known for all interns in this program. It sure is a record for me. And it was good timing because my parents are here this weekend, they came down on Saturday. So it was great.

I just came back from medical records at Mount Scopus. I had a lot of charts to complete and I had to go over today because they were threatening to withhold my paycheck if I didn't finish them. It's impossible to get over to the record room during the week when you're on the other campus. It's so weird: I went back to these charts, two of them for babies who died. One was on a baby that was born premature in the bed. When the baby had been born, I was the first one to get to the labor room, and I didn't know what to do. Here's this little chart and I'm supposed to fill out the discharge summary. There was the autopsy report; it said the baby weighed 460 grams and had atelectasis [**collapsed alveoli, the air sacs in the lungs**], visceral congestion [**accumulation of blood in the circulation going to the internal organs**], PDA [**patent ductus arteriousus, a persistence of the opening of the structure that, in fetal life, shunts blood from the underdeveloped lungs to the rest of the body**], and patent foramen ovale [a communication between the left and right atria of the heart that allows oxygenated and deoxygenated blood to mix]; no big deal. It was just a really preemie baby. Four hundred sixty grams, it might just have been SGA [small for gestational age]. At any rate, I still feel kind of sad about it. Maybe we could have done something, we could have resuscitated it. Months later and I'm still wondering. Well, anyway, it's sad. The chart was wafer thin. There wasn't much to it; a baby who was born and then died. All it had was a heartbeat; it never had an Apgar of more than one. [Apgar: a scoring system used in the immediate newborn period designed to measure neonatal well-being. The baby is evaluated in five categories: heart rate, respiratory rate, color, muscle tone, and response to stimuli. Zero, one, or two points are awarded in each category, and the maximum score is ten. Babies are evaluated at one and five minutes after birth. Apgar scores above seven are considered normal. Less than five are definitely abnormal.] It's very sad. The mother was a thirty-one-year-old woman, she has three living children, maybe she was happy this happened. Maybe eight months from now I'll get named in a lawsuit. Who knows?

You can't focus on the negative all the time, you know? That's what I'm trying to do now, that's my new approach. I still haven't found a way to focus on the positive. In fact, I'm still finding it hard to decide what is positive. I guess it's positive when things turn out well, and when you have a good relationship with a family. That's good. That's the most positive thing I can think of right now.

It's pretty nice having my parents here this weekend. My father rented this total junker of a car. It sounded like the wheels were about to fall off at any moment. It's tiny, you can park it anywhere. I took them down to Soho yesterday; that was fun, they had a good time. Then we went to the Upper East Side and had dinner at this really nice Italian restaurant; nothing fancy, just very nice.

Then I went to sleep and Karen called at some point, I don't know, I was half asleep, I don't remember anything she said. I've had so many conversations with Karen over the phone where I don't remember anything, or I remember hardly anything. It's disgusting, it's totally disgusting.

I found myself reassuring my parents last night that this neighborhood was really okay. Isn't that funny? Because just a few months ago I was telling them how much I hated it. I don't like it, really, but I don't feel threatened so much anymore. I don't think it's a really bad neighborhood; it's not beautiful, but . . .

I'm feeling better now, I'm not feeling depressed, I'm starting to feel like there's an end in sight. I know it's too early to be saying that, but if you look at it, there's twelve months in this year. A hundred divided by 12: That's $8\frac{1}{3}$ percent of the internship per month. At the end of this month, I'll have completed four months; that's $33\frac{1}{3}$ percent. If you add in vacation time, well, then you're at $41\frac{2}{3}$ percent. So at the end of this month, if you include vacation time, which is really free time, I'll have completed over 40 percent of my internship. Not bad! Not bad at all! Forty percent: That has a definite, hefty ring to it. Forty percent! That means I have survived the beginning. I really have, I've gotten through the hell of being an early intern. Whatever hell lies ahead, and I'm sure there's more, at least this has been survived.

Sunday, October 13, 1985, Late at Night

My parents have gone. I felt kind of lonely after they left. I sat down and did a little paperwork, paid a bill, wrote a letter to somebody, and I feel a little better now. It bummed me out because I have to go into work tomorrow and be on call. And I know I need a vacation, I know I need it, but I'll just have to wait. I know I'll make it.

> *I'll just have to wait.*
> *It'll be so great.*
> *To be on vacation*
> *And to sleep late.*

Friday, October 18, 1985

Postcall. Cleared my bed for roaches. In bed here, getting ready to make the big snooze after a rough night on call at University Hospital. Tomorrow I've got to decide what to do with my life. I've been talking with the director of the program I originally wanted to go to up in Boston. He says there's a place for me back there for next year if I want it. I have to decide if I should stay or if I should go. I don't know what the fuck to do, and I'm too tired to think about it now.

Yesterday I got my first kid with AIDS; actually it's ARC **[AIDS-related complex]**, but still, it's the first kid I ever admitted with the big "A." I don't know, it's no big deal; it's just another horrible, fatal disease. Our team got a talk from the immunologist about the disease. Part of it was about health workers with AIDS, and every one of us started wondering if we had it. We were wondering whether we, the people who suck meconium out of the mouths of newborns have gotten AIDS yet. Who knows? Sometimes I think maybe I should go and get myself tested or something. Then at other times I think, what difference does it make?

I had dinner tonight with Ellen. Bought a little Indian food and came back here and ate it. She conked out and went home. That's life when your only friends are interns.

Monday, October 21, 1985

I didn't get much sleep last night, less than an hour. It was a pretty hard night at University. Goddamn renal transplant patient came in! Nice kid for a whining three-year-old. Got his mom's kidney. Jesus Christ, I think the nephrology attendings sit around just thinking of more tests they can order. Anyway, they transplanted the kidney and the kid looked like a million bucks afterward. I hadn't slept all day, and I looked like about thirty cents! I spent two fucking hours in the recovery room; it seemed like I was in there forever. Before the kid came in, there were like a thousand people, a big commotion, everybody wanted to get involved. Then when they saw nothing too exciting was happening, they all split and all of a sudden it was me sitting there alone. All the nurses, all the nephrologists, all the surgeons, they were gone! There's just this kid and I'm in charge. It turned out to be no big deal; nothing happened. I had to make a couple of decisions, but hey, I think I know how to do that. But I didn't get much sleep.

And so what do I do? I get out of work, come home, and stay awake for like three hours! I could have gone to sleep at like seven or eight o'clock. I'm totally overtired. I slept through half of attending rounds today. I didn't even make any bones about it, I just leaned on my elbow and went to sleep. I must be crazy!

I don't know, I guess I just like to come home and pretend I have a life or something. So you know what I wound up doing tonight? I watched TV. I haven't sat and watched TV since well before my internship. Isn't that interesting? Very interesting. Interesting as a pond of mud.

Tomorrow's Tuesday, and Mike Miller is probably going to ask me what I'm planning to do next year. I don't know what the fuck I'm going to tell him. I don't know if I'm going to tell him the truth about this job offer in Boston or what. I'm still trying to make up my mind about it. It depends a lot on Karen, too. She has to decide whether she wants to do her internship in Boston or come down here to one of the real hot shit New York programs. I tried to call her tonight to talk to her about it, but she wasn't home.

My eyes are burning. Someone told me at work that they're really red. What do you expect? Christ, I lie down, trying to get an hour of sleep, and the nurses wake me up. Fucking transplant kid had a headache. I had a headache, too, for God's sake, and I didn't see why the hell they had to wake me up about the transplant kid. But no, they said he had a headache and they took his pressure and it had decided to hop up from 120 to 180.

So I called the renal fellow and told him the kid's pressure was up and he told me to give him a dose of captopril [an antihypertensive medication]. I gave it, the pressure immediately came down, and I went back to bed. About a half hour later, just as I was getting into some deep sleep, a nurse knocks on the door and tells me this other renal kid had a headache. It's an epidemic, for Christ's sake.

I went to see this second patient, and she said that not only did she have a headache but she also had fucking blurry vision. And her blood pressure was 200. So I gave her some captopril, too, and her pressure came down, but she was still complaining of the headache and the blurry vision. I thought about it for a while and I figured, screw it; if her pressure's down, there's nothing wrong. There wasn't anything else I could do.

I went in to talk to her and tried to calm her down. She said she'd try going to sleep. I went back to the on-call room but now I couldn't fall asleep. I was too worried about her. So I got back up and went to her room and sure enough, she was sound asleep.

I kept getting woken up all night long for little things. It was a quiet night and I still couldn't get any sleep.

Another week to go at University Hospital. It'll be a great feeling to be a third of the way through the year, knowing I've completed four tedious months of internship.

Friday, October 25, 1985

I just woke up. I was on call the night before last and I'm on again tonight. I'm doing an every-other, which is okay, I guess, because I get the weekend off. Thank God.

Karen called last night, I think. I think she called and said she'd been offered a place in Boston at the program she wanted to go to. But I'm not sure. I was so tired! I've managed to hold them off here. Mike Miller asked me whether I'm planning to be a resident here next year. He offered me the job. I told him I wasn't sure yet, that I had to do some thinking. We've got to make a decision about this pretty soon.

Tomorrow is my last day at University Hospital. So what have I learned this month? I don't know. Maybe I've learned how to handle many patients all at once. I don't think I learned too much about kidneys, even though that's about all we see.

I better get up before I fall back to sleep.

Saturday, October 26, 1985

My last night at University turned out to be pretty shitty. Everybody on the ward was sick. They all had fever spikes and high blood pressure and headaches, and everything necessary to ruin my day.

And then in the evening, Henry got sick. I guess I haven't mentioned Henry yet. He's an eighteen-year-old with Down's syndrome who's had end-stage renal failure for a long time. He's pretty high-functioning: He's no genius or anything, but he's a really sweet kid with a good personality, and everybody loves him.

Henry has a cadaveric transplant [**transplanted kidney obtained from a cadaver**] and hasn't been doing very well. He came in at the beginning of the month with rejection crisis. We gave him steroids and he got better and went home. Then earlier this week he got admitted because he was rejecting the kidney again; he's been hypertensive and peeing tea-colored urine for days. We've been giving him these massive doses of antirejection medication but he seems to be getting sicker and sicker.

He got really sick last night. He'd been feverish in the afternoon, but we weren't told about it. The nurse who took his temperature called the renal fellow directly and didn't bother informing us. At about six o'clock last night, the fellow called to find out what the results of the sepsis workup were. "What sepsis workup?" I asked. That's how I found out that Henry had spiked a fever.

I went in to see him at that point. He was feverish to about 102 and was tachypneic [**breathing rapidly**]. He didn't look very comfortable. I ordered some Tylenol for him and told the nurses to sponge him down. As the night wore on, he became more and more tachypneic and his fever just wouldn't stay down.

At about ten o'clock, he was looking really uncomfortable. I ordered a chest X ray and drew a blood gas. I brought the blood gas down to the lab myself but I couldn't find the technician for about ten minutes. He was hiding in a back room somewhere. I finally found him and he ran the test grudgingly. The gas just showed a little hypoxia [**the oxygen was a little low**], so I went back upstairs, put Henry in 35 percent oxygen, and then went down to radiology to look at the chest X ray. It didn't look too bad, but when I got back to Henry's room, he was looking more uncomfortable. We turned up his oxygen to 50 percent and, after about another half hour, I did another blood gas. I ran it down to the lab, and again I couldn't fucking find the technician anywhere. I looked all over the damned place; the guy was nowhere to be seen. I spent at least fifteen minutes looking for him. And when I finally found him and told him that I had a really sick patient and needed the gas stat, he said, "I have something else to do stat. I'll get around to yours when I have a chance."

I was ready to strangle the guy! I had already wasted fifteen minutes looking for him and now he was telling me I was going to have to wait longer. And for all I knew, while I was down fucking around with this technician, Henry could have been arresting up on the ward. I was tempted to run the blood gas myself, but I knew that if I touched the machine the guy would have my head on a platter. And yelling at him was completely pointless because I knew that the more you yelled, the more hostile he'd become, and the longer he would take to run the sample. So I just sat on the stool feeling my blood pressure go through the roof.

Finally, the guy picked up the sample, strolled over to the machine, and did it. The gas was pretty lousy. I ran upstairs and showed the resident. Henry was going into respiratory failure in front of our eyes. He was just going down the tubes. And he was scared shitless.

We called the nephrologist and told him we were going to intubate Henry and bring him over to the ICU, and he told us to go ahead. We called the anesthesiology resident, and he did a great job of intubating him. We put him on a respirator and brought him down the hall [**University Hospital's ICU, located down the hall from the Pediatrics ward, admits patients of all ages**].

I spent the whole night in there with him. He was all squared away by about three in the morning, but we had to stay and monitor him constantly. At least he was stable on the respirator. He wasn't getting any sicker and he seemed more comfortable. And I didn't have to deal with that lab technician again because the unit has its own blood gas machine.

Then, this morning, we spent until about noon rounding. I felt like warmed-over shit. After rounds, I went back to the ICU to write a note in Henry's chart, detailing what had happened the night before, and as I was sitting there, one of the nurses called a code. There was this obese, middle-aged guy whose heart had completely stopped beating. He was in fucking asystole [**without a heartbeat**]! And there was no doctor around other than me.

I jumped up onto his chest and started doing compressions. After about ten minutes, the critical-care fellow showed up with a bunch of medical residents trailing behind. But all those guys did was stand around for a while and discuss what to do. They didn't even offer to help. Finally, they decided to shock him [**apply an electrical charge to the chest in hopes that this will start the heart beating again**], so I jumped off his chest. They got his heart beating again and they figured he was stable, so everybody disappeared. I sat back down and tried to finish my note, but fifteen minutes later the same nurse yelled out that the guy had arrested again. And again, nobody was around. So I jumped back on his chest and started CPR again. This time the critical-care fellow showed up with the code team really fast, but again, they all stood around talking while I was doing the resuscitation. Finally I said, "Does somebody else want to do this? I'm from Pediatrics, for God's sake!" I couldn't believe it.

Finally, one of them took over for me and I went to finish my note. They got the guy's heart beating again, but what would have happened if I hadn't been there? The guy would have fucking died because there wasn't a doctor around to do CPR. And I wasn't even supposed to be there. What kind of care are these patients getting?

Amy

Saturday, October 5, 1985

I really needed things to calm down a little after last month ended, but it doesn't look like that's going to happen. I finally got away from Barry Bresnan, but I wound up on the Adolescents ward at Mount Scopus, and it's been terrible. I'm tired, and I'm fed up. I need a vacation. I have one coming up at the end of this month. I hope I'll be able to make it till then.

It's been really busy on Adolescents and it's very depressing. Half the beds are filled with fourteen- and fifteen-year-olds with leukemia, brain tumors, you name it. And the other half is all girls with anorexia nervosa who are completely crazy! It's impossible to talk to them, it's impossible to do anything for them. They just want to be left alone, do their aerobics, vomit in any secret hiding place they can find, and lose weight. They don't want anyone coming near them, especially not a doctor who might actually be able to do something to help them.

My most difficult patient is this eighteen-year-old girl
with choriocarcinoma [**a malignant tumor derived from
the products of a pregnancy**]. Two weeks ago, she was
completely fine. She knew she was pregnant, and she was
looking forward to having the baby. Then suddenly she
started having some cramping and bleeding and she came to
the ER where the gynecologist saw her. They figured she was
having a miscarriage so they did a D and C. The stuff they
took out looked strange. They sent it to pathology and found
out it was cancer. She got admitted to me on Monday, the
first day of the month. We did a quick workup that showed
the cancer had spread all over the place, to her lungs, her
brain, everywhere.

On Wednesday or maybe it was Thursday, she got sick;
she had a headache and was vomiting. The oncologists figured it must have been due to mets in her brain [**metastases
to the brain cause increased pressure within the skull,
leading to symptoms such as headache, vomiting, irri-
tability, etc.**]. We started chemo [**chemotherapy**] on Thursday afternoon, but that didn't help. In fact, she felt a hundred
times worse; she spent all yesterday vomiting. It's pathetic.

I hate going into her room. On the one hand, I know her
prognosis isn't bad, even with the mets; but on the other
hand, I know what this girl's going to have to go through
over the next few months: She's going to have a lot of
chemo, her hair's going to fall out, she's going to be vomiting constantly, she'll have to spend a lot of time in the hospital. Also, if we can't eliminate the cancer with chemo, the
next step is to do a hysterectomy. It's really sad. And so sudden; I mean, one day, she was looking forward to having a
baby, the next day she finds out that not only is she having a
miscarriage, but also she's got cancer that's spread all over her

body, and if it can't be controlled with medication, she may have to have a hysterectomy and never be able to have any children. She's very depressed; you can't blame her. I haven't told her about Sarah. I don't know if it would be good for her to know I've got a baby. It might depress her more.

She's the worst, but they're all like that. I've got an eighteen-year-old who's got a brain tumor. He was fine until about six months ago, when he started waking up having to vomit every morning, with really bad headaches. He came to the emergency room; they did a CT scan, and it showed this huge mass. He's spent a lot of time in the hospital, getting the works, surgery, chemotherapy, radiation, but he's just gotten sicker and sicker. At this point he knows he's going to die and he doesn't want anything done anymore. He screams at any doctor who comes near him. He only trusts one person and that's one of the nurses.

The only good thing about working on this ward is that there're a lot of doctor types around, so nobody gets too many patients. We've got three interns and two subinterns on our team and the whole ward holds only thirty-five patients, so the most anyone can get is seven if things are equally distributed. But seven of these patients are worth fourteen University Hospital patients!

There is one other good thing about this ward: Susannah's working here this month, too. If she hadn't been around last month, I'm positive I would have gone crazy. As it was, I don't know how I survived it. So it's good to have her here. But I can see that neither of us is going to be able to keep the same hours we had last month. The days of getting out at three o'clock in the afternoon are definitely over.

Things at home are quiet. Sarah has a little cold, but I don't think it's too bad. I'm not worried about it. I've been feeling kind of sick myself, so I think we've probably got the same virus. I spent most of today sleeping. I'm on tomorrow, so I won't get to spend any time with her again. It's really impossible being a mother like this.

Thursday, October 10, 1985, 9:00 P.M.

What a week! This has definitely been the worst week of my internship. Every bad thing you could think of has happened to me. And it's not over yet. I've got all day tomorrow and I'm on Saturday.

First, Sarah got sick. She woke up last Saturday night at about 3:00 A.M., screaming her brains out. She had a fever of 103 and she just wouldn't stop crying. I was sure she had meningitis; that's what kept going through my head, "She's got meningitis, she needs a spinal tap, and she'll have to be hospitalized for two weeks and then she's going to end up retarded and I'm going to have to go to work tomorrow and every day from now on and try not to think about it." I somehow got her and myself dressed and we brought her over to the Jonas Bronck emergency room. She kept crying all the way over; she was inconsolable. Rhonda Bennett was the night float. She saw Sarah right away. I was sure she was going to examine her and turn to me and say she needed to draw blood and do an LP **[lumbar puncture]**, but no, she found that all it was was a bad otitis **[otitis media: infection of the middle ear, a common medical problem during the first few years of life]**. She just gave us some Amoxicillin **[an antibiotic used to treat otitis media]** and Tylenol and said she'd be fine in a couple of days.

What a relief! We went back home but I didn't sleep a wink for the rest of that night. Sarah stayed up screaming until about five in the morning, when I guess the Tylenol started to work. She finally fell asleep, but I had to get up and go to the hospital and be on call like usual, as if nothing had happened. I was tired and nauseous and I couldn't concentrate on anything all day except Sarah. I called Larry about forty times. Sarah had a fever most of the day but it was gone by the night and Larry said she looked better.

I admitted this girl with conjunctivitis [**inflammation of the conjunctiva, the outer part of the eye; also called "pink eye"**] who turned out to have GC [**gonococcus, the bacteria that cause gonorrhea; gonococcal conjunctivitis can be very serious, potentially causing blindness, therefore the infection must be treated very vigorously**]. I washed my hands about a hundred times after I examined her but I was sure I got some of the stuff into my eyes. I was positive I was going to turn out to have GC conjunctivitis.

I got maybe an hour of sleep Sunday night and then I had to stay to start fresh on Monday. By that morning I was sick as a dog. I was vomiting and I was sure I had a fever. Larry was going to take the day off from work and stay home with Sarah, but he got called into his office on some emergency, so he had to leave her with Marie, who I'm not sure I trust with medication. I tried to get my work done so I could get home, but I was running at about 10 percent of my usual speed. Susannah told me just to go home, that she'd take care of my patients, but she wasn't feeling so well herself. Of course, nobody else volunteered to help, neither of the subinterns, who have a total of three patients between them; or the other intern, who I must say had to go to clinic that afternoon; or the resident. I plugged on and on and I didn't get home until about seven. By that point I felt like I was going to die. I didn't even care about Sarah anymore. I just got into bed and fell asleep.

When I woke up on Tuesday my eyes were glued shut. I had conjunctivitis. I was sure I had GC. I wanted to go to the hospital and scratch that girl's eyes out, but when I tried to get up I realized I wasn't going anywhere. I literally couldn't get out of bed. My arms felt like they weighed a ton each. I called Arlene, the chief who was on that day, and told her I had the flu and wasn't able to come in. She gave me a really hard time. "Are you sure you can't make it?" she asked me. "Are you really too sick to come to work?" I couldn't believe it!

Larry had to go to work again that day. Before he left, I had him go out and get me some medicine and I wound up staying home and fighting with Marie all day. We mostly fought over little things, but I'll tell you, it's a good thing I'm away at work all day because if I had to spend a lot of time with that woman, I'm sure it'd be the beginning of World War III. Of course, if I wasn't away at work all day, there'd be no reason for her to be coming into my house.

I couldn't face spending another day alone with her, so I went to work yesterday. I was also on call, and there's no way you can call in sick on a day you're on call. When I got to the ward, I found out nothing had been done on any of my patients the whole day before. They had left everything for me! It wasn't like everyone was busy or anything like that, they just didn't think to help me out. So I had a stack of labs to check, notes to write, consults to call, tests to schedule, tons of scut. I got yelled at by the oncologists for not doing something on one of their patients. How could I have done it? I was sick in bed all day!

And I found out that my patient with choriocarcinoma had developed a painful infection in her mouth. She got these sores last week and they were getting worse so I sent off a culture and it turned out to be monilia [a **common type of fungus**]. It was so painful, she hadn't been able to eat or drink for thirty-six hours! She couldn't chew or bite, and nobody had even thought to give her anything to make her feel better. When I went into her room, she was literally crying in pain. I talked to the oncologists, and they hadn't even been told about it. We decided to give her some morphine to see if that relieved the pain. I shot it into her IV and held her hand. Within three minutes, she was feeling better. What's going to happen to her when I leave at the end of the month?

Anyway, I didn't have a bad night, but I only got about three hours of sleep, and this morning I was feeling lousy again. I don't give a damn anymore about anything! When everybody showed up for work this morning, I just told them I was going home. I came home, went to sleep, and slept until a little while ago. Larry's home and Sarah's pretty much back to normal. I feel better myself today. It's not good being a sick intern. You never get a chance to recover. I can see that this flu is going to drag on until the end of the month because I'm just not going to get a chance to rest it out until then. I'm tired and sick, and I'm upset that I'm not able to be a mother to my baby. I really need this vacation. But even that's been a headache.

We're planning to go to Israel at the end of the month. I don't know if I've mentioned that Larry's whole family lives in Israel. His sister is getting married and we're going to the wedding. But Larry's mother gave me a list of things she wants me to bring for her; nothing expensive or anything, just things that are hard to get there. In order to get the stuff, I'd have to spend a whole day shopping. I don't have time for that. I can't just take a day off from work to go shopping! So I told Larry that unless he goes out and gets the things himself, we're just not going to be able to bring it with us. I'm sorry, but I don't think it was right of her to ask me to get it. It goes back to the fact that she doesn't understand what this internship business is all about. Nobody understands it. Nobody could possibly know what it's like to go through this unless you've gone through it yourself.

The other problem about going on vacation is my schedule. We've got tickets to leave from Kennedy International Airport on Monday, October 28, at 11:00 A.M. My vacation officially starts that morning, but of course I'm on call the night before. If I work that night, as I'm supposed to, there's no way I'll be able to make it out of the hospital before nine o'clock in the morning. And considering that I'll be signing out to interns who have never been on the ward before and don't know any of the patients, it'll probably be much later than that. There's no way I'm going to make it to the airport on time. So I'm trying to make a change. The only person I could possibly switch with is one of the subinterns. She says she'll think about it, but she's not committing herself yet. If she can't do it, I just might not show up that day.

Sunday, October 20, 1985

I got into a big fight with Marie on Wednesday. I was postcall and tired but I knew we needed some things, so I called before leaving the hospital and told her I was going to be stopping at the supermarket and was there anything special we needed? She said there wasn't, so I just got the things I knew we needed. Then when I got home, I found out we were out of Pampers. I went crazy! "Didn't you realize we were out of diapers?" I asked. She said she forgot, and I let her have it! The bottom line was that the baby had to spend the night without a diaper change because I was too tired to go out again.

Then Larry yelled at me for yelling at Marie, and I let him have it, too. I told him I wasn't a superwoman, I couldn't do everything. I can't be expected to work and take care of the baby and do the shopping and the cooking. I told him he was going to have to pitch in more and do some of the things that needed to be done around the house. Then after I let him have it, I felt even worse because I hate fighting with him. He's one of those people who just doesn't fight back; he never seems to get mad. He's so calm, it drives me crazy! So we've been fighting a cold war since then.

And things are terrible at work. I hate my patients, except Lisa [**the young woman with choriocarcinoma**]; I hate the house staff I'm working with, except for Susannah, who's had the flu herself; but most of all, I hate the medical students. They're terrible. I've never seen anything like it. I've got this third-year student who's been assigned to me the past two weeks. He actually deserves to fail. He's an M.D.-Ph.D. [**a special track in which college graduates are accepted for a course of study that will ultimately grant then both an M.D. and a Ph.D. degree; Schweitzer accepts a handful of these students each year**], and he's got this attitude problem. He thinks he doesn't have to do anything, all he has to do is show up. I ask him to go and check labs and do other scut and he actually refuses. He says it's not in his job description.

And the subinterns are big pains also. I asked one to switch with me so I don't have to be on the last night of the month. At first she said she'd think about it. Then a few days later she said she'd do it. Then this week she told me she'd decided she couldn't switch for some reason. She didn't tell me why, but it doesn't matter. It was the one time this whole year I really needed somebody to help me out, and nobody would do it. So now I don't know what I'm going to do. Maybe I'll just walk out of the hospital early the next morning. I mean, I'm not going to miss my flight.

My patient with the brain tumor died last week. He had been in a coma for about a week. After talking it over with his parents, we all agreed to make him a DNR. I don't think he was in much pain; his heart just stopped one night and they declared him dead. I never got a chance to talk to him. He was always so angry when he was conscious.

At least Lisa's doing much better. I'm glad about that. She's the only patient I care about right now. Her mouth sores are better and she's able to eat again. She finished her first course of chemo and she's not nauseous anymore. She's actually improving; she's about the only patient all month who has gotten better.

I told Lisa about Sarah on Friday. She asked me if I was married and I told her about Larry and the baby. She said she was happy for me and that she was really sorry that she wasn't still pregnant because she really wanted a baby of her own. She said she thought I must be a good mother because I'm a doctor and I know what to do when children got sick. I don't know about that, but it was a nice thing for her to say. She said she'd like to meet Sarah sometime, so maybe tomorrow, when I'm on call, I'll have Larry bring her over to the hospital.

Well, a week from tonight I'll be on call for the last time on Adolescents. And a week from tomorrow, we'll be on our way to Israel. I can't wait.

Saturday, October 26, 1985

I can talk for only a minute. I'm very busy tonight, but I wanted to get this down on tape before I left on vacation. This is a terrible story.

Tuesday I admitted this seventeen-year-old named Wayne who'd had leukemia in the past but had been in remission for years. He came to the hospital because of shortness of breath, and as soon as he hit the floor, it was pretty clear that his leukemia had recurred. His shortness of breath was due to his enormous spleen. His white count **[while blood cell count in the peripheral circulation]** was over a hundred thousand **[normal is between five thousand and ten thousand]**, and he was anemic and thrombocytopenic **[thrombocytopenia: low platelet count; platelets are factors that aid in clotting of blood]**. The hematologists jumped on him right away. They gave him all sorts of poisons to bring his white count down. He was sick, but he was in pretty good spirits, considering what was happening.

Then yesterday morning, we were on work rounds and Wayne's mother came to tell me he was acting funny. I didn't think much of it, but I went to check him anyway. He was acting really strangely. He was shifting around in bed making gurgling sounds; he didn't respond to questions; it was like he was in a coma, but his eyes were open and he was moving around. I got him to respond to pain, but he didn't respond to anything else. I called for help, and everybody came running. The resident noticed that his right pupil was fixed and dilated **[a sign of an acute and serious change in neurologic function]**. Then he arrested.

We all worked on him for about an hour. We were never able to get anything back. Everybody was in there: Alex

George [**the director of the intensive-care unit**], the chief residents, everybody.

That was my first death. All I kept thinking about through the whole thing was Sarah. A patient's death always bothers you, but when you've got a baby, it means a lot more. I went home after work and just hugged her and hugged her. She'll never understand it. I don't think I'll ever forget it.

We're in the middle of packing. I've got to get everything done tonight because I'm on call tomorrow and there's no other time to do it. I explained the situation to the chief residents and they told me they were sorry but there was no way they could force anybody to switch with me against their will. They at least said that if everything was under control, I could sign out to the senior who was covering at 7:00 A.M. and leave. I'll still never have enough time to get home. So it looks like I'm going to be going on a twelve-hour plane trip after being on call. If it's a busy night, I may not even get a chance to change my clothes or take a shower. I'm really pissed off.

The fact that nobody'd switch with me when I really needed it has made me incredibly angry. It may be a little thing, but I'll tell you, I'm never going to help anybody around here. Except for Susannah, maybe.

Mark

Friday, October 4, 1985

They're trying to kill me. I know they're trying to kill me. I'm just surprised I've survived this long. I've been at Jonas Bronck since Monday, and so far everyone's tried to kill me, the chief residents, the nurses, the elevator operators, the lab technicians, and especially the patients, but no one's managed to finish me off yet. They're all trying, so I know it's only a matter of time.

See what a vacation will do for you? It clears your head, makes you see things in a new light.

Actually, my vacation wasn't exactly what I'd call wonderful. No, wonderful is definitely not the word I'd use. How would I describe it? What word would I use? Lousy; lousy is definitely a word I'd use, lousy bordering on shitty.

It started off with my brother and me in my car driving south as fast as we could to escape from the Bronx. We didn't have any real end point in mind; I was just trying to reach a place where cockroaches don't exist. Actually, that's not true. We were heading for Cincinnati. We both have friends there, and we decided to go visit them. Yes, there's nothing more romantic than spending a week with your younger brother visiting friends in Cincinnati in late September. The whole experience almost made the Bronx seem nice.

Okay, so it wasn't romantic, but Carole and I sure made up for that in the second week of my vacation. We went to the romance capital of the East, Pocono Castle, a resort hotel catering to the honeymoon crowd. What a place! I knew we had made a big mistake when the first thing the bellboy showed us was the heart-shaped bathtub in our room. Carole said she liked it; I thought I showed great restraint by keeping myself from puking right there on the spot. But that wasn't the worst of it. We went down to the dining room for dinner that first night and discovered that everybody there, every last couple, was there on their honeymoon. It was Carole and me and four hundred newlyweds! The place was disgusting; the food was horrible, the decorating job was ostentatious, the rooms were dirty, and it rained all week. All for two hundred bucks a day! Just the kind of relaxing environment I needed.

In the dining room, they put us at a table with another couple. These two were great: They had just been remarried for the second time. They were reformed drug addicts. We spent every meal chatting about AIDS!

And then we got back to the Bronx last Sunday night. I started on the ward at Jonas Bronck Monday morning and it's been hell, absolute hell! On Monday we started work rounds at eight o'clock and didn't finish until three in the afternoon. Seven solid hours of rounds! I'm sure all our names are going to be listed in the *Guinness Book of World*

Records under the category of "World's Longest Work Rounds." Every time we'd try to leave one patient and move on to the next, some disaster would occur and we'd have to stop and sort things out and then try to start again. I picked up five patients: a pair of twins with AIDS who, if you can believe it, are actually named "Winston" and "Salem" (as a result of taking care of them, I've decided to name my first two kids "Chesterfield" and "Lucky Strike"); a brain-dead kid who inhaled a little too much carbon monoxide when her apartment caught on fire; and the "specialties of the house," a couple of asthmatics. That wouldn't have been so bad, five new patients, but you've got to remember, this was my first day on the inpatient service at Jonas Bronck, and I never got a chance to get myself oriented. I didn't know where the labs were, I didn't know how to get results of anything, it took me a day and a half just to figure out where the damn bathroom was! I had to hold it in for thirty-six hours, which, if you've never done it, is not the most comfortable thing in the world. It got so bad, I started to feel like a water balloon. And to make matters worse, the intern who had been on the ward before me hadn't written off-service notes on any of the patients. How considerate! So even if I had had a chance to read the charts, which I hadn't, I still wouldn't have been able to figure out what was going on with my patients.

Then I was on call that first night. I picked up five additional patients and I didn't get any sleep. And the way the schedule worked out, I was on Monday and again on Wednesday and I was postcall on Tuesday and Thursday, so my brain's been in hyperspace for an entire week. I wasn't sure I could find my way back home today, let alone try to figure out what was going on with my patients.

I've been completely helpless. Mike Miller is my attend-

ing this month, and he and the senior resident have asked me
a lot of questions on rounds and I haven't even been close on
any of them. I feel like a real idiot, which I probably am. And
since Miller's a friend of my family's, I've felt even worse
about it. I mean, I'm sure I've gone a long way to convince
him I got into medical school on the one scholarship given
every year to the most deserving mentally retarded individual
in the United States. I've also had this fantasy that he's been
calling my mother at home every night and telling her what a
moron I turned out to be.

But hey, it hasn't all been my fault. I've gotten some
really sick patients over the past few days, and working in
Jonas Bronck isn't exactly my idea of living in paradise! What
a place! The elevators don't work; people have died waiting
for them, and those weren't patients, those were interns! The
people who work in the labs have a combined IQ of about 3.
If you're nice to them, they'll screw you; if you're not nice to
them, they'll screw you. I'm convinced they sit around up
there trying to come up with the most difficult ways to give
out results. If you call them on the phone they'll put you on
hold for ten minutes and then hang up on you. If you call
back and say you were cut off, they'll yell at you and say it's
not their job to give results over the phone. I walked into
the bacteriology lab Wednesday night at ten o'clock and a
woman pushed me out of the door and said they were closed.
They were closed! What does that mean? You're allowed to
diagnose infectious diseases between the hours of nine and
five only?

And the food stinks! That hellhole downstairs is the worst
excuse for a coffee shop I've ever seen! I was wondering how
they could get away charging only a dollar and a half for a
turkey sandwich, so I tried one and I figured it out: They

don't put any turkey on the bread; they don't sell turkey sandwiches, they sell mock-turkey sandwiches, for God's sake! And the place keeps the same hours as the bacteriology lab, which is pretty telling. So when you're up all night, when you really get hungry, you can't even get a mock-turkey sandwich.

And how can Miller and all those other guys who run this department and who say they're really concerned about our well-being not provide us with a shower? If I don't get to take a shower after a night on call, I'm worthless. I feel like shit! Just working in this hospital, you wind up covered with about an inch and a half of municipal hospital crud, and if you don't get a chance to wash that off, you just can't work effectively.

Well, now I've really depressed myself. I was in a pretty good mood when I started recording this, but listening to what I had to say, I have to admit, I must have been pretty crazy to have come back from vacation. I mean, a week at Pocono Castle was a picnic compared to an afternoon at Jonas Bronck.

Monday, October 7, 1985

I think I've finally recovered and gotten myself a little better oriented to Jonas Bronck. It was pretty rough there for a while. That's a real problem with this program: You work at so many places, by the time you feel comfortable at one hospital, they move you to another and you have to start from scratch again. It's the concept of perpetual confusion, an ancient form of torture developed, I believe, during the Ming dynasty.

Things have calmed down a little on the ward. I was on last Monday and Wednesday and then again on Saturday, but from here on it's every third for the rest of the month. (Oh, what joy, only every third night! How lucky can I get?) All last week was a blur. I admitted about ten patients, some of them really sick. It was interesting. They kind of came in groups: Monday was Leukemia Day. First I admitted this fourteen-year-old boy whose gums had started to bleed a lot when he brushed his teeth. He also had a slight fever, he was feeling kind of sick, and he had bone pain. He turned out to have ALL [**acute lymphocytic leukemia, the most common form of cancer in the pediatric age group**], and he's in the poor-prognosis group for everything—age, sex, race, you name it. He only stayed at Jonas Bronck overnight. On Tuesday we shipped him over to Mount Scopus, where he'll get started on a chemotherapy protocol.

Next, I admitted this eight-year-old who had had ALL diagnosed six years ago. He had been treated with chemotherapy for four years and had been considered cured. He woke up one day last week with a terrible headache. He was seen at every emergency room in the Bronx, and all the doctors told his mother the same thing: There was nothing wrong with him, and she should try giving him some Tylenol. He finally showed up to see his hematologist here at Jonas Bronck, and she noticed that he had a sixth-nerve palsy. [**The sixth cranial nerve, the abducens, runs the longest course in its path from its origin in the brain to its point of action in the eye. Because of its long path, the sixth nerve is sensitive to increased pressure within the skull. If pressure is increased, the sixth nerve will not work properly and is said to be "palsied."**] She arranged a stat CT scan that showed a golf-ball-sized mass in his frontal lobe, most probably a lymphoma but possibly some other terrible horrenderoma. [**I don't think the term "horrenderoma" requires defining.**]

So Monday was a really comforting night. I'm starting on

a new service in a new hospital, and I picked up five patients who didn't have adequate off-service notes, so I didn't know what the hell was going on with any of them, and I admitted two kids with terrible prognoses. And I barely had time to recover from that night when I was on again on Wednesday, and that turned out to be Chromosome Abnormality Night. My first hit was this three-month-old who looked like she'd been dead for a few weeks already except for the fact that she'd just had fresh casts put on her legs to correct her club-feet. Her mother brought her into the ER with a complaint that she was breathing too fast. The intern who saw her in the ER asked if she had any problems with her health, and the mother said, "No, there's nothing wrong with her." Turns out she's got trisomy 18 [**a congenital malformation syndrome caused by an extra chromosome No. 18 in every cell of the body**] and severe congenital heart disease, which had caused her to go into congestive heart failure and led to the breathing problem. [**Almost every child born with trisomy 18 has congenital heart disease. The heart disease is one of the factors that leads to early death in these patients. Eighty percent of trisomy 18 children will die before their first birthday.**] The kid's got trisomy 18, and the mother says there's nothing wrong with her! When the intern got the old chart and found out the diagnosis, the mother still denied it. Sounds like she got some top-notch genetic counseling! Anyway, the kid was in congestive failure, so I started her on some Lasix [**a diuretic that rids the body of excess fluid; since fluid buildup is a major problem in congestive heart failure, treatment with Lasix often relieves symptoms such as rapid breathing**] and suddenly she started looking like a million bucks.

I don't really understand why this kid's got those casts on

her legs. I mean, if she's got such a bad prognosis, what's the use of fixing her clubfeet? It doesn't make much sense. Of course, that doesn't mean anything; there aren't many things around here that seem to make much sense.

Anyway, later on Wednesday night (actually it was about five o'clock on Thursday morning) I admitted a fourteen-year-old with Down's syndrome who was having an asthma attack. Now, asthma's pretty straightforward. At this point, I can manage asthma in my sleep. In fact, that's when I usually do the best job. But this kid was just a touch more complicated because, in addition to his Down's syndrome and his asthma, he also had chronic renal failure. The problem with that is that aminophylline [**a medication used in the treatment of asthma**] is removed from the bloodstream by the kidneys. I could give the stuff to him and it'd probably help his asthma but he wouldn't ever be able to get rid of it. He'd probably wind up with toxic side effects; he'd stop wheezing and start seizing, and that wasn't going to be very helpful. We thought about it for a while and decided to turf him [**turf: internese for "transfer to another service"; also referred to as "a dump"**] immediately to University Hospital, where the renal dialysis unit is located. So I had to ride over to University Hospital in the ambulance with him and drop him off at the ward.

Well, I got over to University, brought the kid up, and introduced him to Andy Baron, who was the intern on call over there. He wasn't exactly happy to see me. What the hell happened to him? I mean, I haven't seen him in a couple of months. He used to be a kind of easygoing, friendly guy. He yelled at me when we got there. He accused me of dumping this kid on him, which of course I was. But, hey, that's not my fault. We weren't doing it to make Andy's life more miserable, we were doing it because it seemed to be the best thing for the patient.

It's kind of scary, running into people you haven't seen for a while. It's like going to see a horror movie and realizing that you and your friends are the main characters. You look at Andy and you see what kind of a monster he got turned into and you start to wonder if maybe the same thing hasn't happened to you, but you haven't noticed it because you live with yourself every day and it's hard to notice any changes. I don't know, it's pretty scary.

Anyway, Saturday was pretty quiet. I got only two admissions the whole day, so I had a chance to spend the day sitting in the residents' room watching TV and eating mock-turkey sandwiches. It was good; I needed a chance to relax a little and figure out who all my patients actually were. And today was pretty quiet, too. So right now I've got most of this straightened out.

I spent yesterday in the city with Carole. We went to a matinee. We saw *Cats*. I fell asleep four times and Carole had to wake me up because my snoring was disturbing all the people around us. I don't think I missed much; from what I saw during the short period when I was conscious, it looked pretty lame. I can't figure out how they get people to wear those silly costumes eight times a week. I guess if they can get me to do the ridiculous stuff I do, they can get anybody to do anything!

Thursday, October 10, 1985

As if my life weren't bad enough, Hanson's back. I couldn't believe it! I got called down to the ER to pick up an admission last night and all I knew was it was supposed to be a six-month-old with diarrhea and dehydration. I got down there and there he was. There was no doubt about it. No other six-month-old I've ever seen has that characteristic putrid look about him. His mother had brought him to the ER, and of course she didn't recognize me. During the entire month he had made my life miserable on Infants', she had never once come to see him. All I knew about her was that she was a drug addict. When I realized she was his mother, I almost said, "Oh, I'm sorry, I didn't recognize you without the needle in your arm," but I showed excellent restraint and tact and kept quiet. I'm getting really good at this patient-relations stuff. Anyway, poor little Hanson had been home maybe two weeks and had developed diarrhea again. I asked his mother why she didn't bring him back to Mount Scopus since, after all, all his records are there, and she said, "I don't like those doctors at Mount Scopus. They don't know nothing. They're all a pain in the ass." I guess she came to Jonas Bronck because she wanted a second opinion. Very smart!

Anyway, I figured I should try to get some history on his last days at Mount Scopus, so today I got in touch with Jennifer Urzo, the intern who discharged him from Infants'. I said, "I just admitted Hanson," and I could hear her groan. It's amazing how one tiny baby can cause such a uniform reaction from everybody who's come in contact with him. Jennifer told me this weird story about how she spent the entire month of September working really hard to get him straightened out. Not only did she work on getting him relatively healthy, she also arranged for a bunch of outpatient follow-up appointments, home visits by the VNS [**Visiting Nurse Service**], social work involvement, the works. It sounded as if she genuinely liked the kid (there's no accounting for taste). When everything was arranged, Jennifer somehow got in touch with the mother and actually convinced her to come in and learn how to take care of the baby, which must have been a major miracle. The mother came a few times. Jennifer said she never felt really comfortable around her but that the nurses felt the mother knew what she was doing. So Jennifer told her she could take the baby home. She started going over all the appointments she had made and the mother said something about the fact that she was going to take the baby home and she'd keep most of the appointments but she sure as hell wasn't going to come back to see Jennifer because she didn't like her, she didn't think she was doing a good job with the baby. She said she didn't think she was a very good doctor. This obviously really hurt Jennifer. She even got tearful on the phone just talking to me about it.

Anyway, he seems to be doing okay now. I kept standing around his bed all night waiting for him to crump but he never did. He hasn't had any diarrhea since early this morning but we're still keeping him NPO. I started an IV last night. What thrilling memories that brought back! I'll tell you one thing: He may be a little bigger and two months older, but he still has shit for veins! I stuck him about a dozen times to get the line to stay in place. It felt so comfortable. Just me and him and a box of twenty-two-gauge plastic IV catheters.

While I had Jennifer on the phone, she told me what finally happened with Fenton, the kid on Infants' whose grandmother was crazy and was getting him admitted to all the hospitals in the area for GE reflux that he didn't really have. BCW took the grandmother to court, and Jennifer had to testify, since she was the discharging intern. The judge decided the grandmother needed psychiatric evaluation and ordered that the kid be placed into foster care until that was completed. So there's another happy ending.

Anyway, there's not much else going on right now. I'm going to sleep so I can be bright-eyed and bushy-tailed for call tomorrow.

Wednesday, October 23, 1985

Nothing much has been going on. The ward's been pretty quiet, thank God, and I've been getting sleep most nights when I've been on. The most notable thing that's been happening is that Carole and I have been talking a lot more about the future, which is starting to scare me to death. She wants to get married. I'm not so sure I do. I have to admit, she's been very good to me since I started this insanity. She's always been there when I've needed her, but marriage, that's a really big step.

I'll have to do a lot of thinking about this. Just as soon as I have enough time.

Bob

In October of my internship, my wife developed severe abdominal pains. She went to a physician who did an upper GI series and diagnosed a gastric ulcer. The doctor started her on Cimetidine, a drug that decreases the amount of acid produced by the stomach, gave her advice regarding her diet, and told her to take it easy. With time, the symptoms disappeared.

I was pretty surprised when Beth told me about her ulcer. After all, I was the one with the stressful life-style; I was the one who wasn't getting enough sleep, who was taking care of critically ill patients and trying to cope with their families. All she was doing was working in her laboratory, the same as she had done for the previous three years.

Looking back on it now, it's clear that Beth's life at that time had become as stressful as mine. First, because I was in Boston and she was in New York, she had become a regular weekend passenger on the Eastern shuttle. Beth was terrified of flying, and these weekly excursions were rapidly taking their toll on her mental health. Second, upon finally reaching

Boston after each of these hair-raising flights, she was finding that I, once a sensitive and loving human being, had been turned into a melancholy, self-centered wretch. She seriously questioned, at least to herself, whether this "new" me was a permanent change or whether it was just a temporary interruption in our relationship. And finally, she was worried about her work; it wasn't going as well as it should. She was expending so much energy worrying about me and our relationship that she just couldn't concentrate on what was happening in the lab. At the time, I didn't understand any of this. That's because I, like most interns, couldn't see past my own problems.

It's not surprising that Beth's ulcer first appeared in October. It's during October that internship begins to take its toll on everyone. To the intern, it's the start of the winter doldrums: The thrill of being a doctor is gone (that usually occurs back in July), the "newness" of the on-call routine has worn thin, and exhaustion has begun to set in; in addition, the intern realizes there's no end in sight.

October also is usually the time when interns lose all contact with friends and relatives. I clearly remember the routine that became established at about this time in my own life. I'd come home after shopping to pick up a pizza, take off all my clothes, climb into bed, and watch reruns of *The Odd Couple* and *The Brady Bunch* while wolfing down dinner. Then, by about eight-thirty or nine, I'd turn out the lights and quickly fall asleep. Unless forced, I would not leave my house; I didn't go out to movies or to dinner with friends or even to the supermarket to buy food. I didn't have the strength, and I didn't have the interest. I just wanted to be left alone.

Andy Baron is taking to internship just like I did. When I visited him on the ward at University Hospital early in the month, he said he couldn't talk to me. Figuring he meant he was too busy to take a break, I suggested we go out for dinner, and he said, "No, you don't understand. I can't talk to you, and I don't know if I'll ever be able to talk to you again. If I think about what's happening to me, I'll start to cry, and once I start crying, I don't think I'll be able to stop."

"So you don't think you ever want to talk to me again?" I asked.

"I'm not sure," he replied, and I could see tears starting to well up in his eyes. "For now, I just want to be left alone. I don't want to have to talk or think about anything."

I told Andy how much he reminded me of myself. I told him about my pizza and *Odd Couple* routine. He smiled at this and said it hadn't gotten that bad for him yet (he said his TV was broken).

To people like Andy, Mark, and Amy, who have long-term relationships, October can be a frightening time for loved ones. Because of the appearance of this first wave of depression, interns become introspective, largely ignoring everyone else; all an intern is concerned about is his or her ability to get enough sleep, enough to eat, and to find some kind of happiness without expending too much physical or emotional energy. And like Beth, this is a time when spouses begin to wonder what the future is going to hold.

That's certainly the case with Carole, Mark Greenberg's significant other. She and Mark have had a relationship that's gone on for years and weathered all sorts of storms. And now, four months into his internship, she finds that Mark has no time for her. He falls asleep whenever they go anywhere together; he talks about nothing but life on the wards while taking no interest in her work or her problems. Carole is looking for some answers: She wants to know if the change she's seen in Mark is permanent or temporary; she wants to know what effect all this will have on her life; and she wants to know if she and Mark will wind up getting married or not. Unfortunately, at the present time, Mark is in no condition to give her these answers.

Amy seems to be handling her life outside of work better than any of the other interns. From what she tells me, her relationship with Larry has not been adversely affected by her internship. I think there are two reasons for this. First, Larry is an exceptional guy; he's very patient and understanding, and he loves to spend time with his daughter, which is pretty important, since he's Sarah's primary caretaker most nights and weekend days. Second, Amy is reacting differently than everyone else; she understands that she can't come home, sit in front of the TV, and tune out life, because she has a second job that's even more important than being an intern: She has to come home and be a functioning and loving mother. So far, Amy and Larry seem to be holding things together.

But even their relationship has clearly been stressed. Amy told me about an argument she and Larry had concerning Amy's yelling at their baby-sitter. She said they almost never argue, but this one had occurred after a particularly bad night on call because Amy was overtired. So even for them, some small cracks have begun to appear in their ironclad marriage.

October also is a pretty hard month for the house officers who don't have "significant others." During the middle of this month, I went to the Recovery Room, a bar across the street from Mount Scopus, with some of the senior residents. Ben King told me he was a little upset because he was planning to go to his ex-girlfriend's wedding this weekend. He and this woman broke up last year after a long-term relationship because she just couldn't take it anymore. She didn't like the idea of spending every night alone. "Even when I was there," Ben told me, "I was only about half there."

Usually, the only other single people unattached house officers come in contact with are other single house officers. There's almost no time for them to hang around places where nonmedical single people congregate, and during what little time is available, there is a great deal of pressure to "succeed." But most potential partners, like Ben's old girlfriend, don't want to put up with the bizarre hours interns and residents are forced to keep. So it becomes almost impossible to develop any kind of meaningful relationship that will last through training. For the men, this is just a major irritation; they figure things will straighten themselves out after their internship and residency ends. But for the women, it's a lot more terrifying: They say they can actually hear those proverbial biological clocks ticking away inside their ovaries, and as time passes, they tend to become more and more fixated on finding Mr. Right.

So the overall effect of all this is that a great many interns and residents feel depressed. This internal turmoil can have far-reaching effects, causing the person to decide to make major changes in his or her life. Some interns decide to leave medicine; some (very few) decide to leave life by committing suicide; others decide to leave their program. This last seems to be happening with Andy.

On the evening of October 14, our phone rang almost as soon as I got home from work. It was Andy; he said he needed to see me right away; it was important, and he couldn't talk about it over the phone. He asked if I could come down to the Bronx immediately.

I was worried. I knew Andy had been depressed, and I also knew that anything was possible. So without losing a minute, I got into my car and headed back to the Bronx.

Andy lives on the twentieth floor of the Mount Scopus apartment tower. He has a studio apartment with a terrific view. From the balcony, looking east, you can see parts of the Bronx, City Island, Long Island Sound, and, in the distance, the lights of Long Island and Westchester County. After showing me around, Andy sat me down and said, "I guess you're wondering what's going on." After I told him that was an understatement, he continued: "You have to promise not to tell what I'm going to say to anybody else, not even your wife. It's very secret and I could be extremely damaged if word leaked out."

After we negotiated a little and I finally promised, he told me: "I've been offered a job at another program."

I must admit it was kind of a letdown. I was expecting something juicier, something like he was having an affair with one of the other interns or possibly even with one of his ex-patients on Adolescents. He went on with the story; he told me that before this year had started he had talked with the director of Pediatrics at Boston Children's, the program he had originally ranked first on his match list last year, and had asked him about the possibility of coming back to Boston as a junior resident next year. The director had told him that no jobs were available at that time, but if one were to open up, Andy would be No. 1 on the list. Then yesterday the guy had called, had told him one of the interns was planning to leave and that Andy could have the position if he still wanted it.

Andy told me he was feeling very conflicted. There were a lot of things to think about. First, in spite of everything that's happened, he thinks our program is good and he's made some close friends; he feels bad about the prospect of leaving. Next, Karen is applying to some of the more impres-

sive New York psychiatric residency programs and has a better than even chance of getting in.

"So stay," I said.

"It's not that easy," he replied. "My family's all in
Boston. And I know I'll get a good education at the other
program. And Karen and I are planning on settling in
Boston. We want to have a family; staying in New York will
just delay that."

I told him it sounded like a tough decision but that it
wasn't a bad situation to be in; I mean, he's going to win
either way. I told him I'd definitely be sorry if he decided to
leave.

We talked for about an hour and a half. Of course, nothing was resolved. I'm not sure what's going to happen to
Andy, but if I had to bet, I'd bet on his going back to Boston.
There's simply more to draw him there at this point than
there is to keep him here.

Every October, all the attendings who have contact with the house
staff meet to discuss the internship group. This meeting
serves two purposes: First, it identifies those people who are
or may soon be having trouble so that some form of intervention can be planned; second, the meeting allows us to
come up with some idea of who will be returning the following year and, more importantly, who will be leaving.

It's rare that all the interns come back as residents. Two
members of each year's incoming group are accepted with
the understanding from the very beginning that they'll be
leaving for residencies in psychiatry, radiology, or other specialties in which a year of pediatric or internal medicine train-

ing is mandatory. A few more, people like Andy Baron, decide to change programs for personal reasons. So during this meeting, Mike Miller tried to get a head count of prospective junior residents.

The meeting this year was interesting. Mark Greenberg and Andy Baron were both viewed as very good. Andy, in fact, is considered by most people to be outstanding, an excellent candidate for one of the program's four chief resident positions. Since he had sworn me to secrecy, I didn't mention a word about his job offer in Boston.

The intern about whom there was the most discussion was Amy. She's apparently made more than a couple of enemies among the attendings. In addition to her problem in the emergency room back in July, she tends to do a lot of little things that get people upset, such as leaving the hospital early; complaining frequently and loudly; and criticizing other house officers, such as Barry Bresnan. I spoke up for her; I said I thought a lot of her complaints were justified and that she shouldn't be condemned for voicing them.

It was ultimately decided that Mike Miller would have a talk with Amy and explain the concerns that had been raised at the meeting. It wasn't thought that anything needed to be done, that her work was certainly good enough to justify being offered a position for the following year. I don't know how Amy is going to respond to this criticism, especially since I'm sure she won't think it's warranted. After all, I don't think it's warranted either; I think she should be commended for doing as good a job as she's done, considering all the pressure on her. And I'm going to tell her that.

Andy

Tuesday, November 5, 1985

I've pretty much decided that I'd like to go back to Boston for next year, but things still are up in the air. It looks like Karen might get accepted into Columbia's psych program, and that's going to be pretty hard to turn down.

This past week has been really, really hard, with this decision hanging over our heads, and we've both been incredibly stressed out. I'm in OPD [the Outpatient Department] now on the Jonas Bronck side, although I spend two days a week here at Mount Scopus for clinic. I've been on call it seems like an inordinate number of times already in the past week; I've already done two every-other-nights and I'm on call again tonight. I'm finding the Jonas Bronck ER a real drag to work in. The nurses are extremely hostile and critical and cold. They're very good nurses, very efficient, and they obviously know what they're doing. They're much better than the nurses in the West Bronx ER, but they all seem to have a chip on their shoulder. I've been told that there's some kind of war going on among them but, hey, you know, that's no excuse. That just makes it a drag for everybody else to work there.

The place is unbelievably busy. I end up getting out at four-thirty in the morning on nights when I'm on call. It's just fucked. You come home, you sleep for three hours, and you're supposed to be back at the hospital for the eight-o'clock teaching conference the next morning. Forget it! It's really unfortunate. I really was looking forward to the Jonas Bronck ER, and I do enjoy the work I do there. The pathology that walks through the door, the patient population, the mix, it's unbelievable; it's fantastic. I'd love it except dealing with all these angry nurses is a real drag!

So far I've been thrown into that fucking asthma room a lot more than I think I should've been. Some of the other interns are going to have to help pitch in with that. **[In the Jonas Bronck ER, all patients with asthma attacks are placed in the asthma room. When things get busy, one house officer, usually an intern, winds up doing nothing but working in the room. That person may see nothing but asthmatics for four or five hours at a stretch.]** It gets really boring in there, seeing the same thing over and over again without a rest. I've already complained about it but I don't think anybody really cares. That's all; I've got to go back to clinic now.

Thursday, November 7, 1985

I'm in the P^2C^2 **[Pediatric Primary Care Center, the pediatric clinic at Jonas Bronck]** conference room waiting for the conference to start. I have to talk quietly or they'll think I'm talking to myself. Nobody else is here yet.

Last night I was so tired, I slept eleven and a half hours straight. I could have slept another five easily. Can't work every-others, they just wear the shit out of you. And on both of those every-others, I worked in the ER till about 5:00 A.M. Then yesterday I had to work in the ER all day, from nine to five. Jesus Christ, this place is a goddamn zoo!

Karen's been here for the past few days. We're still trying to decide whether to go back to Boston or stay in New York. It's tough, there are a lot of things to consider, but so far it looks like we're both leaning toward going home. They've been really good about it here. Miller knows what's going on, and he's giving me the time I need to decide. He says he wants me to stay. It's nice of him to say it, but does he really mean it?

I've got to stop now; someone just came in.

Friday, November 8, 1985

I'm here in the thirteenth-floor conference room of West Bronx, where the pediatric OPD conference is supposed to be. I got here at eight, and I just found out it doesn't start until eight-thirty, so I'm about twenty minutes early now.

Today's an important day because after long and tedious deliberation, Karen and I have definitely decided to go back to Boston for the remainder of my residency. Karen's been getting internship offers from everybody. Every single place she's applied to is offering her a position. It's hard to turn opportunities like that down, but we've decided to go back. Karen feels she'll be happy at Boston University, the program where she'll wind up going, even though it's not in the same league as Cornell and Columbia.

The important thing that we've decided is that we want to be around family and friends. This year has been so hard for me because of the separation. I don't think internship can ever be easy, but I know I would have been better off had I stayed in Boston instead of coming to the Bronx. Plus, we both think Boston is a nicer and easier city to live in than New York, which is very exciting but also crazy and congested and stressful. So this morning I'm going to call up Scott Thomas, the director at Boston Children's, and tell him I'd like that spot if he's still got it. If he hasn't got it, I'm going to wring his little neck. Then I'll have to call Mike Miller and let him know that I've made my decision. Hopefully I'll get the contract and the whole thing will be signed, sealed, and delivered within the next week or so.

A couple of days ago, I thought for the first time that I'd ultimately like to subspecialize. It's not because I have some burning interest in any one field; I don't really. A lot of things interest me, but there's not any one field I'm that attached to or interested in. I want to subspecialize because I'm tired of being so inexpert at so many things. I don't think I could spend the rest of my life knowing a little about a lot of things, like many of the OPD attendings do. I need to feel like there's one area in which I have a great depth of knowledge. Someone just walked in; I'm going to have to stop now.

Sunday, November 10, 1985

Karen's still here. I'm finding myself getting all depressed again, and I'm not really sure why. I just can't put my finger on it. There's just nothing that's obviously wrong: I'm in OPD, and that's pretty easy, I'm not lonely. Something's just wrong.

I guess one of the reasons I'm depressed is that I made this massive decision about going back to Boston, and now there's a kind of letdown. It's official now: I called Dr.

Thomas the other day and accepted the job. And Karen officially turned down Cornell and accepted the place at Boston
University. On Monday, Thomas is going to call Mike Miller
and discuss it with him, then he's going to call me back and
let me know it's all sealed, and that's it! That's it; we're
going back.

Last Friday was a horrendous day. I started working at my
clinic at Mount Scopus at nine o'clock, and I began the day
with a child-abuse case. A patient who requested me because
they had seen me in the emergency room once, came in. I
saw signs of abuse and reported them. That was a very
unpleasant situation; they were very angry, and I can understand why. And then at noon, I went to the Jonas Bronck ER,
and I was there until seven the next morning. I was so tired,
I fell asleep taking a history! While I was talking to the
mother, I just zonked out! Soon as I woke up, I picked it up
with the next question I had in my mind, but I knew I had
been asleep. The mother was sitting there staring at me with
a strange look in her eyes that hadn't been there a second
before (because it probably had been several seconds). Then I
fell asleep listening to some asthmatic's chest several times. I
kept wondering, Why is it taking me so long to get a respiratory rate? Because I kept falling asleep every time I put the
'scope on the kid's chest, that's why! So that was an abysmal
night.

I guess all the shit I've been seeing at Jonas Bronck's
depressing me a little, too. All the child abuse and the codes
and all that, that stuff gets me down. And it's been really
cloudy and nasty and rainy, and that doesn't help. And living
in the Bronx is just a bore.

And there's something else: I've started to become
obsessed that I've got AIDS. I've started waking up in the

morning feeling anxious, thinking I'm going to die. That's one of the main criteria for major depression. I've been trying to go and get the test [**blood test for antibodies to HIV**] but I haven't done it yet, initially because if I think rationally about it, there's no reason I should have it, and then because I realize I don't want to find out if I'm positive.

I'm getting a little bit of the feeling I used to have in medical school, that I'm trapped in a prison, and the rest of the world out there is beautiful and happening and I'm not in it. I saw *The New York Times* today; I read the headline saying that the Democrats had taken over the Senate, controlling fifty-five of the seats. I didn't even know there was an election. I didn't know until after it was over. So I feel very much isolated from the mainstream of humanity. And at times I feel like I'm not taking this seriously enough. I mean, each mother brings her kid in and the kid means all the world to her, but to me, it's just another set of wheezing lungs. I try to do my best with each one, I try to think of each one individually and I do, I know I do, but, I don't know, in some ways it all becomes a blurred mass of humanity flowing through the doors of the emergency room.

There are these two kids, I see them all the time, the mother calls me every week, she comes into clinic every week. She's a really good mom, maybe a little neurotic. She has a Down's baby; the other kid's normal. And she's really great. Seems like there are so few other patients and families I'm happy to see, though. That can't be right; you can't just like one family out of the hundreds who come through.

The streetlights are still on. It's the middle of the day but it's so dark that the lights are still on. I'm supposed to go shopping to get my brother and his wife a wedding present. I don't know what the fuck to get them.

Wednesday, November 13, 1985

It's cold outside, it's turning into winter. You can see your breath in the air. I'm still in the OPD. And I'm feeling better.

My depression has gone, for the most part. At least the acute exacerbation. I'm still left with the chronic, smoldering depression I've had since August. It turns out I was also getting sick. Got this goddamn viral syndrome from some kid and now I've got this residual cough.

Monday, November 25, 1985

I haven't talked into this for a while. Karen left yesterday, and when she's here, I usually talk less. I'd rather spend time with her than this machine.

Today's the end of the fifth month. I finished outpatient this afternoon. Tomorrow I start on 8 East at Jonas Bronck. And while part of me is relieved to get the hell out of that ER, which has just been a madhouse, I'm kind of dreading tomorrow because I'm on call and I have my clinic, so it's going to be a dreadful night. I'll be up all night. I'm already sure of that.

But I'm also looking forward to being back to the somewhat protected environment of a floor where I know what my work is. The work's cut out for you, and even if what most of the other interns have said about Jonas Bronck wards is true, that there are too few nurses up there and the nurses who are there don't want to work, in a lot of ways it's better than being out in the unprotected emergency room.

I've been paying more attention to some of the other interns lately. Some of them are a lot worse off than I am. Take Peter Carson, for instance. I've been working with him in the ER. My God, is he an angry man! He makes the rest of

us seem like laughing hyenas. I'll give you an example. Saturday we were both on call. It was a horrendous day in the
emergency room. The third year resident was Larry Brooks,
and he said it was the worst day he'd ever seen in that ER. It
wasn't because we had so many terrible things happening. We
did have a few kids in the back [**the back: the trauma area of
the Jonas Bronck emergency room**], but there weren't any
real tragedies that took up a lot of time. It was because of the
volume; it just never let up, and there were only three of us
working until four in the afternoon when the evening float
resident showed up four hours late (ooops). I literally had
only ten minutes to eat during the entire nineteen hours I
was there. It was exhausting. By 4:00 A.M. I was just going
cross-eyed. I couldn't concentrate for shit.

Anyway, at 4:00 A.M. we were ready to get out of there,
but the triage box wouldn't empty. Finally it got down to two
charts. The night float was there, he was all alive and peppy,
and we were getting ready to leave, but Larry came in and
told us there were a couple more to see and I heard the night
float say, "Just give them to the interns and go home yourself." Well, when Peter heard that, he went completely
berserk. He started screaming, "That fucker! Let the interns
do it? Let me at him! I'll rip his testicles off, one by one!" He
was screaming so loud that everybody in the emergency
room could hear him. A nurse came knocking on the door in
a second saying, "You know, not everybody out here wants to
hear about testicles being torn off!" But Peter was beyond
help; he was so incensed, he just kept screaming. We were
saying, "Peter, Peter, shut up or we're going to have to call
security on you," and then he kind of calmed down, but only
a little. He was wild. And then what ended up happening was

that Larry told us to go home and he wound up seeing the last patient himself.

Peter and I split a cab back to our apartment building, and all the way there he was just cursing, saying how much he hates being an intern and how much he hates the ER. He was just absolutely infuriated. But he's back there every day, somehow or other. I guess I'm not the worst off, but I think I'm getting a reputation as being one of the depressed interns.

Amy

Saturday, November 16, 1985

I'm not very happy about being back from vacation. We had a wonderful, relaxing time in Israel. I must have been in pretty bad shape before we went away. The frightening thing is I didn't even realize it until I had a chance to get away from it for a while.

Before we left, I was obsessed with being on call that last night of October. It became the most important thing in my life. As it turned out, it wasn't a problem. It was a very bad night; I admitted a new onset diabetic who was in DKA [**diabetic ketoacidosis, a buildup of acid in the blood caused by the inability of the body to use glucose as its energy source; insulin, the protein that allows the blood's glucose to enter the cells of the tissues, is either absent or abnormal in diabetics**], and I was up all night managing the boy's fluids and electrolytes, but Ben King, the senior who was on with me that night, threw me out of the hospital at seven in the morning. Just like that, he told me to leave and have a good time and not to worry about a thing, he'd take care of the patients and sign out to the new interns. So, after all that, I did manage to get home, take a shower, and change my clothes before we had to leave for the airport.

The flights were terrible both there and back. Sarah screamed the whole way. It didn't bother me that much on the way over; I was completely zonked and I slept most of the trip, so it was Larry's problem, not mine. But I couldn't believe it on the way back! I was sure the pilot was going to land and throw us off the plane. But outside of the flights, it was the best vacation of our lives. Larry's parents were great. They wanted to spend the whole time taking care of the baby; they left Larry and me alone, and they encouraged us to go out on our own and do whatever we wanted. I slept late every morning, and by the middle of the second week, I felt like I had finally caught up on my sleep. We traveled all over the country. I can't imagine a better vacation. The only problem was that the time just flew by. Before we knew it, it was time to pack up, get on the plane, and come back to work.

I've been working in the OPD at Mount Scopus. Things have been quiet. I've been getting out between twelve and one on the nights I've been on call. Things would be perfect if we weren't all so jet-lagged. When I'm not on call, I've been going to sleep at seven and waking up at three in the morning. And when I am on call, forget it; I have to use toothpicks to keep my eyelids open after nine. But I can see how much calmer I am now compared with before we left. I really needed that vacation, there's no doubt about it. It's just too bad I have to wait so long for my next one to come around.

Saturday, November 23, 1985, 2:00 P.M.

I had a very bad night last night. The ER was busy and depressing. I didn't get home until after three this morning, and I just woke up about a half hour ago. It's a beautiful day and we're going to take Sarah out for a walk in a few minutes, but I wanted to record this while it was still fresh in my mind.

At about nine o'clock, I picked up the chart of a three-year-old whose mother said she had had a bloody bowel movement earlier that evening. I didn't think much of it at first; bloody bowel movements aren't that unusual. It's usually due to an anal fissure [**a tear in the anal mucosa caused by straining; very common in children around toilet-training age**]. I called the patient in and I saw she was a cute, well-dressed, healthy-looking little girl. I took the history from the mother, who seemed appropriately concerned. Then I examined the girl. I noticed right away that her rectum was very red and it looked kind of . . . well, boggy is the best way to describe it. I did a rectal exam and I noticed that the tone of the sphincter seemed a little decreased. I was suspicious, so I called the attending and did the rest of the rectal exam with him in the room. The girl didn't even cry while I was doing it. There wasn't a peep out of her, which, to say the least, is not normal for a three-year-old.

I got a sinking feeling in my stomach when I was doing the rectal because I've taken care of little kids who've been sexually abused and I knew what was going to happen from here. I was going to have to question the mother, she would probably deny everything and accuse me of making it all up, we'd get into a big fight, and she'd eventually start to cry. Then I would have to call the BCW and report the case to them and they'd wind up doing a full investigation, which might end with them taking the child away from her mother. I knew that none of this was fun or interesting and it was going to take up most of the rest of the night.

Anyway, I started asking all the questions I had to ask. Did they live alone, or were there other people living with them? Did she watch the girl all the time, or did she leave her with other people? Was the girl's father around, and did he have anything to do with her? The mother knew something was up because she answered every question honestly and without too much expression. It turned out that the mother and the girl lived in a two-bedroom apartment with ten other people. Some of the people who lived there were relatives, like the girl's grandmother; some were friends of their family; and some were just friends of the friends. The woman's father had been a junkie and had died of AIDS the year before. The family had all been tested for HIV and the girl's grandmother had been the only other person who tested positive. But some of the other people living in the apartment were junkies, and they hadn't been tested. And there were two teenage boys who were cousins of the mother and who had been taken into the apartment when they themselves had been abused by their own parents a few months before. It was a very confusing, chaotic story, but I believed it because it wasn't all that unusual. I've heard lots of stories like this one since I started medical school.

The woman said she and her daughter slept in the same bed at night but during the day the mother went to school and she had to leave the girl with anybody who happened to be around. She admitted it was possible for anyone, especially her teenage cousins, to have sexually abused the girl while she was out of the house.

At this point the mother started crying and I had to leave the room for a minute. I was ready to cry myself.

I went over to talk to the attending and told him the story and he asked me one question: Why was the mother being so honest? I hadn't even thought about it before that, but he was right; having your child sexually abused by one of your relatives is not something anyone would be especially proud of. The only thing I could think of was maybe the mother wanted to get something out of this. I mean, here she is living with all these people in this chaotic apartment. Maybe she figured the BCW would do their investigation and decide that the girl should stay with the mother but that they should be placed in their own apartment. It was a pretty disgusting thought but completely possible.

I had to go back and tell the mother what was going to happen, and I had to do the rape kit. The attending told me I should draw some blood for HIV testing, just as a baseline. **[People who are exposed to the human immunodeficiency virus will test positive for antibodies to the virus a few weeks after the exposure. As such, Amy's patient should have been negative but may later convert to positivity if she had been exposed to the virus.]** I hadn't even thought about that, but it certainly was a possibility. Not only did this little girl get raped, but also the rapist might have given her AIDS! I didn't even want to think about it.

After I finished the rape kit, I started to make all the phone calls. I first called the social worker, and she said that I'd have to make it a joint response. **[Joint response: When a child's life is considered to be in danger, a report must be made simultaneously to the Bureau of Child Welfare and the New York City Police Department. The BCW's investigation does not get started immediately. Therefore, an immediate investigation by the police must be done to determine whether the child can return home.]**

So I called the BCW and the police. The whole thing, from start to finish, took about four hours. By the time I was done it was after one o'clock in the morning and there were still a bunch of charts in the box. What finally happened was the mother and the girl were placed in a shelter for the night. I think they'll ultimately get placed in their own apartment.

I've been thinking about that little girl constantly since I finished with her. All through the rest of the night, all during the cab ride home, while I was trying to fall asleep and since I woke up, that little girl didn't leave my mind. It's really terrible. I'm sure I'll see her face in front of me for years and years to come.

Mark

Friday, November 1, 1985

Yesterday was Halloween. I was on last night in the Jonas Bronck ER, and I learned an important lesson: If you want people to trust you, it's probably not a good idea to dress up like Bozo the Clown. I know that because I did dress up like Bozo the Clown yesterday and none of the parents of my patients wanted to have anything to do with me. I guess I can't really blame them; it's one thing to come to the ER and wait four hours to be seen by a competent, or at least a semi-competent, doctor. It's another thing to wait four hours and finally get called in to find out your kid's going to be treated by Bozo the Clown.

But, hey, it was Halloween, and we're supposed to be taking care of kids, aren't we? We all decided the day before to come in dressed in costumes. Peter Carson, who's about six feet three and weighs at least 250 pounds, came dressed as a ballerina, the chief residents were dressed as killer bees,

Terry Tanner (a junior resident) was dressed as a witch, and I was Bozo the Clown. The kids seemed to like it even though their parents weren't ecstatic about it. And everything would have been fine if I hadn't had to tell a mother that her kid was dead.

It was about eight o'clock, right in the middle of the busiest time of the evening, of course. All bad things seem to happen when we're really busy. We got a call from the EMS **[Emergency Medical Service]** saying they were bringing in a traumatic arrest **[a patient who, as a result of some accident, was not breathing and whose heart was not beating]**. So Bozo the Clown; the six-foot, three-inch prima ballerina; the witch; and one of the killer bees stood around the trauma area waiting for the disaster to show up. It took maybe two minutes and they brought in this eight-year-old. He had run out into the street and had gotten flattened by a van. The van then stopped and the kid got pinned under the back wheels. They started CPR out on the street, but you could tell it wasn't doing him much good. He was pulseless and breathless, and when they hooked him up to a monitor, he was flatline **[he had no electrical activity in his heart]**.

We knew it was probably going to be pretty hopeless, but we started doing everything anyway. The chief intubated him, Peter started pumping his chest, and Terry and I tried to get lines **[IVs]** into him. I somehow got one in his right arm, which was a miracle in itself, and we started pushing bicarb **[sodium bicarbonate, to reverse the buildup of acid in the blood]** and epi **[epinephrine, in an attempt to get the heart to start beating again],** but nothing happened. Then the surgeons came by and offered to crack the kid's chest for us [perform an emergency thoracotomy, an operation in which the chest is opened so that the heart can be directly massaged]. Hey, when five surgeons walk up to you with scalpels in their hands and say they'd like to crack a patient's chest, it's hard to say no. So the kid got his chest cracked and they found that he had a bronchopleural fistula. [The impact of the van had caused the left main stem bronchus, the main windpipe to the left lung, to tear in half. Oxygen that was being forced into the boy's windpipe was ending up in the pleural space outside the lung, causing an ever-worsening tension pneumothorax.] It was about then that we realized that this code was pretty much over.

I walked out of the trauma area, and the boy's mother was standing there less than ten feet away. She was literally being held up by one of the nurses. She said, "Doctor, how is he? Is he going to be okay?" I didn't see any way out; I was too upset to come up with a lie. So that's when I, wearing my Bozo wig; my Bozo makeup; my Bozo shoes; and my Bozo suit, which was now covered with blood, told the woman that her son had died.

She went crazy. She started crying and she fell down on the floor. I felt like a total idiot standing there dressed like that, and there was nothing I could do to change anything. One of the hospital administrators, the guy we call the administrator-in-charge-of-patients-dying because he always seems to show up when this kind of thing happens, came, and he, the nurse, and I lifted the mother up off the floor. The administrator led the woman out of the ER. I don't know where they went, but before seeing the next patient, I changed my clothes and took off the stupid makeup. I don't think I'll wear that costume again. The bloodstains kind of take all the fun out of it. And next year, if I'm on call on Halloween again, I don't think I'm going to dress up.

Tuesday, November 5, 1985, 10:00 P.M.

I'm feeling much better today. Sure, a weekend off, that's just what I needed. It gave Carole and me two whole days to fight about whether we should get married. It was a whole lot of fun. At least she didn't make me wear my Bozo costume while we argued.

I really don't know what to do about this. I don't want to get married during my internship. Can you imagine that? Falling asleep standing up right in the middle of the ceremony. And then the wedding night! Yes, the wedding night must be a memorable event for the intern's spouse. Eight continuous hours of snoring. Seriously, being married is hard enough when you lead a relatively normal life. I don't think it'd be possible for us to survive if we got married while I was doing this. But Carole thinks we should do it. She says if we got married, she'd be able to take better care of me for the rest of the year. I think I'll eventually wind up marrying Carole. We get along very well and we basically want the same things out of life. I just can't do it yet. I think I'll be able to think a little more clearly after this is over, but that's not for seven months yet. Well, what can I do? I'll just try to hold her off as long as possible.

And then, after that fabulous weekend, I got to be on call again last night in the Jonas Bronck emergency room. And what a night it was! We were five hours behind the whole time. We had two security guards stationed at the doors to protect us. Every five minutes, another angry customer would appear and want to know why his or her precious little child who had been sneezing for three days hadn't been seen yet. And what interesting patients I had to take care of! I got this four-day-old who, through some sort of screwup in our world-renowned Social Service Department, wound up getting discharged from the nursery with his psychotic mother who also happened to be a crackhead. Usually when a baby's born to a crackhead mother, Social Services picks up their hot line and gets a BCW hold slapped on the kid so that the baby can be kept in the hospital while the BCW figures out what to do with him. We usually have to keep them longer than three days anyway because the kids usually have withdrawal symptoms and need to be treated. But somehow Social Services missed the boat and sent him home early.

When the nursery's social worker realized the kid had been discharged, she called the cops and had them find the kid and bring him back. The cops did a great job: They went out and scooped up the baby, the mother, and, lucky for us, the father and brought them all in. Of course, they didn't mention to them what was going on. So not only did we have two psychotic crackheads roaming around the ER, we also had two psychotic crackheads who were paranoid and had no idea what was going on, which is a wonderful combination.

Well, it didn't take long for them to figure it all out. Once they woke up to the fact that we were planning to admit the kid and slap a BCW hold on him and that their chance of ever getting him back again was about the same as my chance of being elected president of the United States, they let their best qualities come to the surface. The father picked up the baby like he was a football and started to move toward the exit. The city cops, who were still hanging around, knew the mistake had been the hospital's and not the parents'. They also knew that the parents hadn't done anything wrong, at least not at that exact moment, and so they didn't try to stop them. The cops left, and that's when the hospital security guards stepped in. They caught the guy, brought him back, put them all in a room, and watched them for the rest of the night, but they weren't exactly happy to do it. They acted as if they had better things to do than baby-sit for a couple of ranting junkies.

I tried my hand at talking the parents down. I told them that putting the baby in the hospital was the best thing for him, that if he was hooked we could give him medication to make him better and slowly wean him off. It sounded great to me, but of course the parents, who were pretty crazed, didn't buy it. Then the social worker showed up and she talked to them for about a half hour. Obviously she made just as great an impact on them as I did, because when they came out, the mother was still holding on to the baby. We didn't know what to do next, so we had a priest come down and talk to them, we had some friend of the family's who had shown up talk to them, but nothing seemed to do any good. Finally, after about five hours of this nonsense, the mother said she had to get home right away because she needed something to steady her nerves. She handed the baby over to the social worker and she and the father kind of ran for the door. So we got the baby back. Her fix was more important than the baby in the long run.

We got out this morning at four-thirty. I didn't get home until after five, and I fell right to sleep. I didn't even get out of my clothes. I had two and a half hours of sleep. But it was quality sleep, so that makes a big difference. Yeah, right! And when I woke up, I was still wearing my smelly, dirty clothes. What a wonderful experience this internship is!

At least I got to go to my grandmother's for dinner

tonight. My grandmother's good, she's doing fine. I'm sure she thinks I've lost my mind or something because I can't keep up even boring conversations with her and I keep falling asleep every five minutes. But she doesn't say anything. She just keeps the food coming.

I'm going to sleep now. Maybe when I wake up, I'll realize this has all been a dream.

Monday, November 11, 1985

I'm suddenly not feeling very well. I think I'm coming down with something. I've had this stomach ache and a sore throat since this morning. I have the chills, too, so I probably have a fever. I can't understand why I'm getting sick. After all, all I do is hang around an emergency room, working twenty-hour shifts, seeing sick children who sneeze on me, cough on me, pee on me, shit on me, and vomit on me. What possible means would I have of getting sick?

My mother came to visit on Saturday. She walked into my apartment, took one look, and said something like, "Oh, dear, I hadn't heard anything about a nuclear attack in this part of the Bronx." (I get my sense of humor from my mother's side of the family.) I have to admit, I have kind of neglected the housework over the past few months. So my mother rolled up her sleeves, got to work, and spent the next six hours cleaning up my apartment. I had all sorts of great things planned; I was going to take her to lunch at the Jonas Bronck coffee shop. I figured she'd love those mock-turkey sandwiches. Oh, what the hell! We did go out for dinner at a nice Italian restaurant. It was nice to see her. And now I can be sick in a nice, clean apartment.

I took some Tylenol, but it hasn't done any good. I think I'm really sick.

Wednesday, November 13, 1985, 9:00 P.M.

I'm dying. I didn't expect it to be one of the patients who would finally get me, and I never thought they'd use germ warfare. But there it was, the most virulent GI [**gastrointestinal**] bug ever to exist, and now I'm sure it's only a matter of time.

I fell asleep Monday night at about seven. I wasn't feeling well Tuesday, but I made it to work and somehow I made it through the day. I took my temperature in the ER at about one in the afternoon. It was over a hundred. But hey, I'm an intern, and interns can do anything, including working a full fourteen-hour day when they're sick. I came home yesterday and fell asleep right away. I slept until 11 P.M., and when I woke up, I had the worst cramps in the history of the human race. I ran to the bathroom and stayed there for the next four hours. I got back into bed sometime after three and I fell asleep for a while. Then I woke up with worse cramps than before and tried to get up to run to the bathroom. My brain was strongly in favor of the idea but my body just wouldn't budge. I managed to crawl out of bed and make it to the bathroom just in time and I fell asleep in there until my alarm went off at seven-thirty. I still could barely move. At that point, something told me that I probably wasn't going to be able to make it to work.

I called Jon Golden [**one of the chief residents**] and told him what was going on. I told him I was on call and that I probably wouldn't be able to make it in. Calling in sick on the day you're on call is the biggest sin an intern can commit. But what could I do? I couldn't even walk! Jon told me not to worry, just to try to get well and make it back tomorrow.

I got into bed and fell asleep, but Elizabeth woke me up at about ten. She's on the Jonas Bronck wards this month. She'd heard I was sick and wanted to know if I was making it up just to cash in on some sympathy. I guess when she heard my voice, she realized I was serious. She asked if there was anything I needed; I asked for cyanide. She said she'd see what she could do. She asked if I thought I was dehydrated and I said I was easily about 10 percent dry. [**The main complication of gastroenteritis is dehydration. Five percent dehydration is enough to require hospitalization; 10 percent is serious, and 15 percent may lead to shock.**] She said I should come in and let her start an IV. I told her I hated pain, and knowing her technical skills, I would never allow her anywhere near me with a needle in her hand. She thanked me for the vote of confidence and said I must be feeling better to be making jokes. I said, "Who's making jokes?"

I am actually feeling better tonight, but I still fall down every time I try to get out of bed, and I don't think that's normal. I feel guilty about not going to work. I know the other people on call tonight are probably working their butts off and cursing me every chance they get. But what can I do?

This is such a screwed-up job. In what other profession would you actually feel guilty calling in sick when you really are sick? Lawyers get sick and take a week off, schoolteachers take days off like it's coming to them, which it is. It's only us interns and residents who feel guilty about it.

Sunday, November 17, 1985

I'm feeling 100 percent better. Well, maybe not 100 percent, maybe only 80 percent, but I'm feeling well enough today to go into the city with Carole to see a movie. I don't know what we're going to see yet, and it doesn't really matter. The only thing I care about is that the theater has seats comfortable enough for me to fall asleep in. Of course, after last night, a chair with spikes coming out of it would be comfortable enough for me to fall asleep in.

I won't say last night was bad but at about two o'clock there were still about ten charts in the box, and Peter Carson and I were seeing patients in the asthma room. Peter's kid was a really cute three-year-old girl. She was sitting there alone because her mother was out registering her and she was scared. But very shyly, she asked Peter if he was a doctor. "Yup," Peter said.

"And you don't ever go to sleep?" the girl asked.

"Nope," Peter answered.

"Never?" the kid asked, amazed.

"Never," Peter answered. And he meant it. Then we both ran out of the asthma room and cracked up. It seemed really true last night. I've got to get out of this ER!

Thursday, November 21, 1985

Did I say something the other day about having to get out of the ER? Well, the chief residents must have overheard me, because on Tuesday afternoon they called me into their office and told me that I wouldn't have to work in the ER Tuesday night. That would have been wonderful, except for the fact that they also told me that I had been selected, out of the entire intern group, to have the distinction of being the first person ever to take call in the new neonatal intensive-care unit that had opened that morning on 7 South. What a thrill that was! It sure is something I'll never forget as long as I live! And you know what? After one night on 7 South, I'm ready to spend the rest of the year in the ER.

What happened was, Val Saunders was supposed to be on that night, but she called in sick. Everybody said she wasn't really sick, she just didn't want to be on call the first night in the new unit, which was really very sweet of her if it was in fact true. Somebody had to work there, and since I was supposed to be on in the ER and there were four other people down there that night, the chiefs figured I was "it."

So there I was, sitting in the nurses' station on 7 South, waiting for some disaster to happen. I'd never even been in a NICU before, and here I was, taking care of twenty-eight tiny babies. Just looking at them scared me to death! And nobody knew how anything worked! They hadn't even figured out how to turn the heat on yet! They couldn't find the outlets to plug in the damn ventilators! And sometime during the move, somebody had misplaced the coffee pot, so we couldn't even make coffee! And I was supposed to cover all those babies! It's the kind of situation that'd make for an outstanding horror movie!

Somehow, the babies and I all made it through. I didn't get any sleep. I spent the whole night running from room to room trying to figure out what was going on, but nobody died, nobody even crumped, and when the morning came, I was still standing on my feet. I think I did pretty well. Maybe I'll go into neonatology.

By the way, after I finished yesterday morning, I went into the chiefs' office and they thanked me for filling in. I very graciously told them that if they ever pulled anything like that on me again, I'd reach down their throats, yank their spleens out through their mouths, and refeed it to them. I was very polite about it. I think they got the point.

Bob

When I was a house officer, we occasionally saw children who were beaten or molested by their parents or other adults, but these cases seemed to be few and far between. I vividly remember one of the first abuse cases in which I was involved. In clinic one day during a month of OPD at Jonas Bronck, I found that I had been assigned a new patient, a little eleven-year-old girl named Brenda. As soon as I called her into the examining room, her mother began to tell me that Brenda never seemed to have any energy; she was always tired, was complaining of too many bellyaches, and seemed to be gaining a great deal of weight. When I started to examine the girl, it took less than a minute to diagnose the problem: Brenda was about six months pregnant.

Realizing that I now was going to have the gargantuan job of informing this mother that her daughter, a child herself, was going to have a baby, I told Brenda to get dressed and sent her out to the waiting room. I started the conversation awkwardly, asking if Brenda had begun to get her period yet and whether she had a boyfriend. Finally, after beating around the bush long enough, I blurted out the news to the woman.

She wasn't surprised. She told me she had seen her daughter's clothes get tight and had noticed that her breasts had become swollen. Then she told me an amazing piece of news: She knew who the father of the baby was. He was Brenda's fifteen-year-old brother, the person who was entrusted with caring for the girl after school when the mother was at work. The brother, apparently fed up with having been saddled with the responsibility of looking after his sister when he'd rather have been out with his friends, had taken his frustrations out on the girl.

I spent hours with that family. We talked about abortion, an option Brenda's mother rejected for religious reasons. We talked about the effect the pregnancy would have on Brenda and on her brother. We talked about what measures should be taken to prevent anything like this from happening again. Brenda's mother assured me that she would discipline the boy in her own way and begged me not to report the case to the Bureau of Child Welfare, the state-run agency charged with investigating possible cases of child abuse. After long discussions with the mother and the clinic attending, I decided to go along with her wishes. That might have been a mistake: I never saw Brenda or her mother again.

I think I made that mistake for the same reason I can remember Brenda so distinctly: Hers was one of the few cases of child abuse I was called on to manage during my residency. Now, several years later, an intern can't even make it through a single week in the OPD without getting involved with the BCW. Child abuse and neglect have definitely increased over the past few years. Rarely a day goes by now without a family of two or three or four kids who have been abandoned or beaten or sexually molested being escorted into the emergency room. I was working in the ER last week and the cops brought in a family of seven children ranging in age from ten months to eight years. The parents were both crack dealers; the mother had been arrested the previous Thursday, and the father had been taken into custody the day after that. These children had been left to fend for themselves for three full days. They were starving; dirty; and very, very scared. Medical care in their cases included food, baths, and hugs. The police had been called by a neighbor who had complained that the baby was crying too loudly. They were temporarily placed in a shelter.

Although child abuse is clearly on the rise in New York, there's another reason that so many more abuse patients are being identified. The house officers are far more sensitized to the signs and symptoms of child abuse than my fellow residents and I were in the early 1980s. Interns are asking questions I never would have even thought of asking, such as: Where does the child sleep? How many people sleep in the same room? Who watches the child during the day? And they're performing more pelvic and rectal exams in younger kids than we performed. Through these means, they're finding evidence that we simply would have missed.

The net effect of all of this is that the Bureau of Child Welfare has become completely overwhelmed. The BCW always has had its problems. Calling their twenty-four-hour hot-line number to report a family has always been an exercise in frustration. They've been slow-moving in completing investigations. But this has all become worse since the current "epidemic" of child abuse hit. And new methods of guaranteeing the safety of at-risk children have had to be invented.

The "joint response" for reporting serious abuse involving children was only recently developed. When the examining doctor believes that a child has been abused and that his
or her life may be in danger, both the BCW and the New
York City Police must be informed immediately. A member
of the sex crimes unit of the police force is then immediately
dispatched to investigate the situation. The child, who may
have been beaten by the mother's ex-boyfriend, may have
been sexually molested by a relative who lives in the same
apartment, or may have been removed from a home in which
another child has been killed or seriously injured, cannot be
released into the custody of the parents or other relatives
until the results of the police investigation are known. Since
most abused children appear in the emergency room during
the evening or night, sleeping accommodations for the child
must be arranged. This entails either admission to the hospital or, if possible, transfer to some shelter.

To the house officer, dealing with child abuse translates
into pure aggravation. There's endless scut that must be
done. In cases in which sexual abuse is suspected, a "rape kit"
must be completed and followed to the letter so that the collected specimens can be used later in court as evidence; reams
of medical and legal forms must be completed according to
strict guidelines; telephone calls to agents who are themselves
overworked and who aren't always the most caring or sympathetic individuals must be made; and careful explanations to
hostile, suspicious, and often guilt-ridden parents must be
given. All this must be carried out by doctors to whom child
abuse is a particular anathema; these people, who become
adjusted to death and disease, frequently become physically ill
themselves while working with a child who has been abused.

The net effect of all this is that progress through the pile of ER charts is dramatically slowed. A house officer can be tied up for an entire night reporting a single child abuse case. In an emergency room in which three or four residents are seeing all patients, the loss of one or sometimes two doctors can add endless hours to the waiting time. Parents sitting with sick children become angry and hostile as the clock ticks on. Often the whole situation ends with hospital guards being called to protect everyone from injury.

Andy

Wednesday, November 27, 1985

So I survived my first night on call at Jonas Bronck. It was busy, another night of sleep deprivation. Harvey Abelson announced in front of about six people that I was going back to Boston. Not in a real obvious way; no, he was very subtle. He said something like, "So I hear you're going to Boston next year!" He said it really loud, in kind of a nasty tone. So now I feel a little bit like a *persona non grata* around some people. What do people think when you leave a program? Do they think you're turning your nose up at it and, in a sense, at them? Can't they accept that you're leaving because there's something else, something more important than being in this program for you? Why can't they just accept that?

Friday, November 29, 1985

I'm on a flight to Boston, to spent the weekend at home. This is the first year I was away from my family on Thanksgiving. It was a real bummer. I was on call yesterday, and there was nothing to eat. Stupidly, I forgot to bring anything to the hospital from home. I should have known that there would have been nothing available to eat at Jonas Bronck, but I just didn't think about it. I mean, there was *nothing*. I starved during Thanksgiving. What an image! Well, it'll make going home even better. I don't know . . . I just hope I'm never on call for another Thanksgiving.

It's been an amazing few days. I started working on 8 East this week. It took only a few days for word to get around that I'm leaving the program next year, and I've already noticed a big change in the way people are reacting to me. It's a funny thing; some people want everybody to know that I'm leaving. For example, Alan Cozza, the director of the pediatric service at Jonas Bronck, has referred to my leaving in front of a lot of people on a couple of occasions over the past week. He's not doing it with any kind of malice. I'm not really sure why he's doing it, whether there's a certain sense of pride he feels, like he's proud of this program and thinks that when I go to Boston I should reflect how good it is, or if he really wants to mark me as different from everybody else. It's a strange thing.

Thursday, December 12, 1985, 11:00 P.M.

I don't think I've put much of an entry into this thing for a long time. I think I've made it out of my month or two of depression, and I haven't felt the need to vent about things as badly. Even though I'll still wind up cursing about life, and I still hate being on call more than anything else I can ever remember hating, and I'm still chronically tired, I'm definitely not depressed anymore. I don't know why; maybe it's because I know vacation's coming up very soon. Maybe it's because I'm so goddamned used to working all the time now. Maybe it's because I like working on 8 East because of the social feel of the place; it's as if the whole staff is part of one big family. I just don't know.

I find myself looking back on the past six months and realizing that so far this is the month I've enjoyed the most out of all of them. Jonas Bronck and my month at University Hospital are definitely my two favorites. Isn't that strange? University Hospital was like a torture chamber half the time. But there were some really nice things about it, too. Cute nurses, that helped, but I think it was also because the patients were complicated and interesting. Or maybe it was just because it seemed like it was just us residents against everybody else. I think I like the idea of working in a tertiary-care hospital. Everything is right there, there aren't any interruptions from the emergency room or the clinics or anyplace else; it gives you a sense of self-containment. But I also like Jonas Bronck, which is just the opposite. It's hard to justify that, it's hard to figure how I can like two such different systems.

I have to present chief of service rounds again tomorrow. **[These rounds are held every Friday at noon at Jonas Bronck Hospital. An interesting case is selected, a summary is prepared and presented by the responsible intern, and the patient is discussed in depth by the faculty's expert in that particular field. It's a well-liked teaching conference, but it's a pain in the ass for the overworked intern who has to prepare the presentation.]** I wish they'd ask somebody else. It takes a long time to get that stuff together. I did it once already this month and I had to do a grand rounds, too. **[Grand rounds, held on Wednesday at noon, are occasionally constructed around an in-house** patient. As in the case of chief of service rounds, the intern caring for the patient is responsible for preparing the case presentation.]

We've got a pretty good team this month. Our senior, Pat Cummings, has turned out to be okay. I actually like him; he's a good resident. He's kind of got a gruff, hard edge to him, but other than that he's a funny guy.

And I like my medical student. We spent a fair amount of time together today, doing scut. Makes all the difference in the world to do scut with another person. We had to stick some kid five times to get his IV in. I finally got it in, and it ran like a dream. I hope the damned kid doesn't kick it out tonight. It looked good enough to stay in a couple of days.

Last night wasn't too bad. I got four admissions and I managed to get a couple of hours of sleep. And I got everything done. For me that's good. I couldn't do that at the beginning of the year. If I got four admissions back in August, I'd be up all night, writing and writing and writing. Now I write less and go to sleep. By the end of the year, I should be able to do ten or twelve a night. Ten or twelve admissions a night—boy, is that a horrible thought.

I've got the FIB service [**FIB: fever in baby**]. I have six patients and they have a combined age of about nine months. And they all look and act alike; it's hard to tell one apart from all the others. It's not very interesting.

I ran into Mike Miller the other day. When he saw me, he kind of frowned. And I sort of frowned when I saw him frown. I said, "What's the face?" He said, "Well, I'm just sad you're going to be leaving and you're not going to be around here next year." I don't know if that's what he was really frowning about. At any rate, it was a nice thing to say.

This tape recorder's kind of annoying me; it's making

weird noises. It's hypnotizing me . . . Well, you get the idea. Those sleeping noises go on for the remainder of this tape, which I'll be recording over now, because ten minutes of sleeping noises aren't very exciting. I fell asleep again, fell asleep while recording on this fucking tape recorder.

Sunday, December 22, 1985

I've actually wanted to talk into this machine for a few days but I ran out of tapes and haven't had a chance to get any until now. It wasn't that I was saving up anything much to say, just the sense of having something to say.

I've had this whole weekend off and in some ways it felt like my vacation actually started yesterday. I'll be on call tomorrow and then I leave on Tuesday morning. I'm tempted to wear a big button to work tomorrow that says, "THEY CAN'T HURT ME NOW!" I've told several people that no one's died on me yet this month, and everyone's said the same thing: "Don't be so smug. You still have one more night!" But I feel somewhat confident that I'll make it unscathed off this ward and that I'll always have fond memories of general pediatrics when it's provided in a place like Jonas Bronck, surrounded by lots of smart, nice people in a great environment without disasters.

Well, that's not exactly true; I did have one near disaster, a kid with asthma who I was supposed to admit to the ward but who wound up going to the ICU on an Isuprel drip [**Isuprel drip: a continuous IV infusion of isoproterenol, a drug used in cases of asthma when there is danger of respiratory failure**]. I was the one who started that Isuprel drip. I didn't know how to do that before. Hey! I now know how to do an Isuprel drip! Now I pretty much know everything.

We're going to New Orleans this vacation, I just found

out yesterday. I got home post-call, turned on my answering machine, and there was a message from my brother saying he and his wife and Karen all decided it would be great fun to go to New Orleans and they asked if I wanted to do it and I said sure. I'd go anywhere. Just to get the hell out of the damn Bronx!

I know this sounds weird, but a lot of people I talk to say I'm the most enthusiastic person about the program. Isn't that ironic? Here I am, I've been depressed for months, and now I've even decided to leave at the end of the year, and I'm the most enthusiastic of the interns! But it's true, I have been in a good mood for the past month, and I've started to wonder why. I think it had a lot to do with two things: first, being on 8 East, which was really great; and second, knowing I've finally got this vacation coming up. I survived, I've made it through six long months! I've reached the halfway point.

I feel like I've gotten a lot out of this first half of the year. I think I've learned a lot. I don't know how I'll compare with those second years at Children's when I start out there next year. Will they be way ahead of me? Will they know a lot more, having been in that highly academic environment for their internships? Will my vast ability to do scut really pay off at all? Will it matter? I don't know.

Will my learned ability to manipulate the ancillary personnel to get patient care done quickly and efficiently make any difference in a place where the ancillary services are actually good? I mean, I've gotten good at working through this system, I've finally learned how to get things done fast. I've just watched how the third years do it and I've figured out you either stretch the truth or you simply lie outright. You have to make everything seem like an incredible emergency or people will ignore you. You tell the elevator operator that you need an elevator right away or else the patient's going to die. And they'll do it! That elevator will be there in a second. It's too bad, but it's just the way it is here, it's just a game you have to play if you want to get things done or you want to take proper care of your patients. You have to lie; they just don't give a shit any way else.

I'm learning to be efficient and how to be smart. Friday night on call was pretty quiet, I didn't have a bad night, I only got two admissions; one was a FIB, the other a UTI [**urinary tract infection**]. I even got a couple of hours of sleep. So on Saturday morning when I was postcall, instead of going right home, I sat down and wrote four of my off-service notes on my chronic patients. I put them in my clipboard, I'll bring them back tomorrow, date them, then I'll stick them in the chart, and I'll be done with them. The fact of the matter is, I've got to get out of there Tuesday morning and catch a plane, and I want to be able to bolt at early as I can.

The UTI I admitted was kind of interesting. It was a one-month-old who came up as a FIB. Of course, no one in the ER had done a sed rate [**erythrocyte sedimentation rate—the rate at which red blood cells settle when left to stand in a capillary tube; an elevated sed rate is a sign of inflammation and therefore an indication of infection**] or a UA [**urinalysis**]; thanks a lot. It's always the second and the third years who send them in unworked up. So I basically did the whole admission by myself. Pat kind of danced in for a minute, copied down the history I took, poked and prodded the kid, then went back to sleep, and I finished the rest of the workup. I actually got one of the night-shift nurses to hold the kid while I drew the blood and did the suprapubic [**bladder tap**], and there they were on the unspun urine, sheets of polys [**polys: polymorphonuclear leukocytes,**

white blood cells that flock to the site of a bacterial infec-
tion] and gram-negative rods [**the microscopic appearance
of *E. coli***]. So I decided what to do: I wrote the orders to
start the antibiotics, went in, woke up Pat, and said, "Pat, this
kid's got a UTI and I'm starting her on ampicillin and
Cephotaxime [**two types of antibiotics**], a hundred per kilo
of both. How's that sound?" He mumbled something like,
"Huh? Fine," and went back to sleep. In the morning he said
he was very impressed with the gram stain. I had done the
right things. So how do you like that? I can now manage
unbelievably simple problems all by myself!

You know, I might have been able to do all this a few
months ago, but I sure wouldn't have felt right about it. Now
I feel like I know what I'm doing. Watch, I'll come back and
something horrible will have happened by Monday.

I played Santa Claus on the ward the other day. I put on
the red suit and the hat and the big beard and the hair and
stuffed a pillow under there and strapped it on with some
kerlix [**rolls of gauze bandages**] and said, "Ho, ho, ho, ho,
ho! Merry Christmas!" about three hundred times. I danced
around on my toes. Everybody told me what a wonderful
Santa Claus I was. And John Mason [**a four-year-old boy
who is a long-term occupant of the AIDS section of 8
West**], who usually runs away from Santa Claus, thought it
was a lot of fun this year and was totally thrilled. Filled with
happiness. I was glad I could do that, make little John happy
for a while.

Amy

Sunday, December 1, 1985

This internship is really rotten. There's nothing about it that I like. The only thing that gives me any kind of enjoyment right now is Sarah, and I can't even really enjoy being with her because in the back of my mind I always know that I'm going to have to be back at the hospital soon. And when I'm in the hospital, I spend my time with all these nice children who have terrible things wrong with them, and I know I'm going to have to watch them get sicker and sicker for thirty-six hours at a stretch. I don't know, I have pictures of about a half dozen of these children burned into my mind. I haven't been able to forget the little girl I saw in the ER last week who was sexually abused; I've been thinking about her constantly. There's her, that leukemic who died on Adolescent, and a couple of others.

I've been thinking this weekend that if I knew then what I know now, I never would have done this internship. Everything's so hopeless; I'm so hopeless. It isn't even half over yet, I just got back from vacation, and I'm already so tired of all of it! I don't know what I'm going to do. I don't know what I can do.

Mike Miller called me into his office last week. He told me some of the people in the department were kind of upset with my attitude. Screw them! He said there were some people who thought I wasn't taking the job seriously enough. Terrific, just terrific! I asked Mike how many of these people had to take night call every third night with a baby at home. How many of them had to neglect their responsibilities as a parent in order to work in the hospital? I told him I was doing the best I could and if he or anybody else didn't like it, he should just fire me! He said it wasn't him, that he understood what was going on, but that this had come up at some meeting they had had and he had been the one who was supposed to have a talk with me about it.

He also asked if I wanted to come back as a junior resident next year. He said he needed to know within a few days. I told him I'd have to think about it. After what he'd said to me I was pretty damned angry and I seriously thought of going in and telling him to go fuck himself, that I wouldn't be coming back next year or ever again. But after I cooled off a little, I finally told him on Friday that I would be back. I don't know. I could have taken a year or two off; Larry was encouraging me to do whatever I thought was best, and for a while, taking some time off made a lot of sense. But then, after I'd been working in the emergency room for a while, I figured this wasn't too bad and I thought I could stick it out. I still have the feeling that if I were to decide to take some time off, I'd have a lot of trouble getting myself motivated enough to get back into it. But this whole weekend, I've been thinking, maybe I made a mistake. Maybe staying isn't such a good idea. The problem is, I don't know what I can do about it now.

I'm doing Children's at Mount Scopus this month. It was good to get out of the Jonas Bronck ER. The actual work there isn't so bad; it's just that there are so many bad things that happen to the kids who come in, like that sexually abused girl. Seeing kids like that every day gets to be too much pretty quickly.

Coming to Mount Scopus isn't great either. It seems like everybody has an attitude, all the nurses and the clerks. They all seem to resent having us around, like we're getting in the way of their work or something. This week, Harrison [**Harrison Boyd, the other intern on the Northwest 5 Children's team**] wrote a q4h order for Demerol [**an order for the pain killer Demerol to be given every four hours, around the clock**] for this five-year-old with sickle-cell disease who was in with a painful crisis. One of the nurses came up to him while we were on work rounds and said, "We don't give pain meds q4h on this floor." Just like that; she simply refused to do it. Harrison told her that the child was in a lot of pain and he thought she required medication around the clock. The nurse told him it didn't matter, the rules are that they can only give pain meds on a prn basis [**prn: as needed**]. So now this little girl has to feel pain and beg for her medication before someone will give it to her. The nurses say they're afraid a q4h order will make the patient addicted. They say they're afraid they'll wind up turning the child into a drug addict. Turn a five-year-old into a drug addict? That's a lot of nonsense. But that's the way things are done at Mount Scopus, and there's nothing you can do to change it.

There aren't a lot of patients on the ward right now, and most of the ones who are there are just post-op cleft lips and palates [**Mount Scopus has a large craniofacial center**]. We do have a couple of fairly sick patients, and one of them is mine. She's a really sad case: She's a ten-year-old with neurofibromatosis. [**This is an inherited disorder in which, for reasons that are not yet clear, the affected individual may develop dark pigmented spots and a variety of tumors of the skin, optic nerve, the brain, and other internal organs. Most people with neurofibromatosis live a normal life; some are severely deformed.**] She was perfectly well until last week, when she started having trouble urinating. Her mother brought her to the ER last Tuesday and they did a KUB [**X ray of the abdomen; called KUB for kidneys, ureters, and bladder**] and found she had a big mass in her pelvis. Then they did a chest X ray and found another tumor in her lung. She was admitted to Children's, and on Thursday she had a CT scan that showed she also had a brain tumor. They're taking her to the OR tomorrow to biopsy the pelvic and the lung tumors, but the oncologists don't think she's going to do very well. It's too bad, too; she seems like she could have been a nice, normal girl.

Wednesday, December 4, 1985

It's getting cold. It's really turning into winter, and I don't know if I'm ready for that to happen yet. I guess I'm kind of depressed and the weather isn't helping things any.

Things on the ward are getting worse. This morning a little girl with sickle-cell disease died while we were on work rounds. She had been admitted a few hours before with what looked like pneumonia. Harrison had brought her up from the emergency room, started her on antibiotics, put her on oxygen, and went on with the rest of his work. At about eight-fifteen, while we were on work rounds, the girl's mother came up to us and said her daughter was breathing funny. We ran into the room and watched as she arrested. We called a code, and people started flying into the room from all over the place. We worked on her for over an hour, but we never got her pulse back. Her mother was hysterical. It was completely unexpected. After it was over, Harrison locked himself into the on-call room and wouldn't come out. It was terrible. And we still have no idea what happened or why it happened.

On Monday I admitted six new patients, which is busy for Children's. Most of them were pre-ops for Tuesday, but they still needed to be worked up and have bloods drawn and all the rest of the endless scut that goes with an admission. I tried to get my student to help, but she simply refused. I don't know what it is with these medical students; when I was doing my clerkships, I'd sooner jump out a window than tell my intern I wasn't going to do the things he or she asked.

Angela, that girl with neurofibromatosis, is not doing well at all. She went to the OR on Monday to have her tumors biopsied. They turned out to be different types of malignancies, which is very unusual. It's also terrible in terms of her prognosis. She spent Monday night in the ICU and then came down to the ward again on Tuesday. The oncologists are talking about what they're going to do with her, it looks like she's going to need both chemo and radiation, but they have to decide which order to do them in and when to start. Nobody's said anything to Angela yet. I think she knows, though. She's pretty smart and she seems very sad.

I went to talk to Mike Miller today about next year. I went up to his office, and when he saw me standing there, he said, "Amy, what's wrong?" I guess my face told him I was upset. I went into his office and we closed the door and I started telling him everything that had been happening, and about five minutes into it I started crying and I just couldn't stop. He held my hand and tried to calm me down, but I just couldn't stop crying. I cried for fifteen, twenty minutes. I knew he was starting to get bored with me so I tried hard to control myself and finally I stopped. I told him I wasn't sure whether I wanted to come back next year. He told me he understood and that whatever I wanted to do would be okay but that I had to sit down and give it some serious thought, taking everything into account before making up my mind. He said he'd hold off submitting my name for reappointment for another few days but I should try to get back to him by sometime next week. Then I left and went back to the ward.

That's the first time that ever happened to me. I'm not a crier. The last time I remember crying is when my mother died. This isn't nearly as bad as that was, so I'm kind of surprised. I don't know, maybe it's because I'm so tired, or because working in the Bronx is so depressing, or maybe it's because I miss Sarah so much and I know I'm missing so much of her infancy. All I know is, if the rest of internship is going to be like this, I'm never going to make it.

Sunday, December 15, 1985

I haven't recorded anything in a while. I've been going through a very difficult time. I think I'm finally out of my depression, though. I don't know why that happened. Nothing's changed; I'm not spending any more time with my family, the work on the ward is just about the same, but for some reason I'm feeling a little better.

Maybe one thing that made me feel a little better is that I

finally made a decision about next year. I went to see Mike Miller again last week and told him I had decided to stay. He seemed relieved to hear it and then he said some nice things to me. He told me he had been very upset after our conversation the day I broke down in his office and that he was glad I had decided to stay because he thought I'd make a good resident. I don't know whether he meant it or not, especially after what he told me about that meeting the faculty had last month, but it still was a nice thing for him to say. It made me feel good to hear it.

Sarah stood up by herself for the first time yesterday. I was on call, of course, so I didn't get to see it, but Larry called me at work to tell me. He said he and Sarah had been sitting on the floor of the living room and at one point she just climbed up the edge of the couch and stood there. She isn't even nine months old yet! She's done everything early; she sat at five months, so I'm not surprised she stood by herself so young. She's like a miracle; it's absolutely amazing just to be around her and watch as she grows and develops. I've been so worried, but she seems to be a normal, well-adjusted little girl. Maybe she's better off with me at work; maybe Marie is doing a better job with her than I would have.

Things at work are about the same. Angela is getting worse; we started her on chemotherapy about a week ago, but it isn't doing any good. She had a seizure last week. We brought her out of it with IV Valium and Dilantin [**an anti-convulsant medication**], but the seizure wasn't a good sign. She had a repeat CT scan a few hours after the seizure and it showed the tumor was larger than it had been before the biopsy. Right after the seizure, her mother, who's been at her bedside ever since Angela got here, told me for the first time that she thought Angela was going to die. Angela's going to start radiation therapy next week. Maybe that'll help; everyone doubts it, though. I hope they're wrong.

Sunday, December 22, 1985

Tomorrow's my last night on Children's. After that I go into the nursery at Jonas Bronck. I'm not looking forward to that. Nursery's the one rotation I'm really frightened of. My month on Children's wasn't all that bad. It would have been great had I not gotten so depressed in the middle of it. Children's is a good place; most of the patients are pretty healthy and almost everyone gets well without much work from us. Angela isn't getting better, though, and we're all getting depressed about that. I guess we've all come to accept the fact that she's going to die sometime soon. I just hope she holds out till Wednesday. I'd really rather not have her die on my time.

Mark

Tuesday, December 3, 1985

Well, I finally figured out how they decide what ward a kid'll go to when he gets admitted to Mount Scopus. If he still sleeps in a crib, they put him on Infants'; if he's out of the crib but hasn't committed any violent crime yet, they put him on Children's; and once he's committed his first violent crime, he automatically goes to Adolescent. Yes, what a wonderful experience it is to take care of ten- and eleven-year-olds wanted by the police in three states for armed robbery.

Okay, so maybe I'm exaggerating, but only a little. What a weird place this Adolescent ward is! As I see it, we've got three groups of patients. The first are those kids with truly medical problems; these are mainly kids with different types of cancers; they really make things pretty sad. The second are the girls with eating disorders like anorexia nervosa and bulimia; what a thrill it is to take care of eighteen-year-old girls who weigh seventy pounds and think they're too fat, and who spend the entire day exercising and trying to find places to hide their vomit. And the third group are the drug addicts who come in for detox and for antibiotics for their skin abscesses. Yes, this is all a very rewarding experience.

And, of course, everybody's got VD. I've had to do pelvic exams on everything that's come through the front door! At least on every female; I haven't had to do one on a male yet, but hey, the month's still young, and anything can happen. It doesn't even matter what they come in with, they always wind up having PID [**pelvic inflammatory disease, the common name for infection of one or more pelvic organs, usually caused by gonococcus, the bacteria that causes gonorrhea; PID is diagnosed by pelvic exam and confirmed by culture results**].

Take this afternoon, for instance. I was on call last night so I wasn't supposed to get any admissions today, but there were a lot of electives scheduled and the other intern was at clinic, so they asked me to take a kid who came in at about three. Her only problem was an ASD [**atrial septal defect, a "hole" in the structure that separates the right and the left atria**], and she was coming in for a cardiac cath. [**This is a test in which a catheter is passed through a blood vessel in the thigh and fed up to the heart. In pediatrics, cardiac caths usually are done on children with ASDs and other congenital malformations of the heart to better define the anatomy in preparation for surgical repair.**] A pretty straightforward case, right? Sure, until I started taking the damn social history. I made the mistake of asking her if she was sexually active, and she said she was. (At least she didn't say, "No, I just sort of lie there." Someone once told me they actually had a patient who said that.) Then I asked if she used any form of contraception and she said she didn't. So I asked if she wanted to be pregnant and she said no, and I started to go into my "If you don't want to be pregnant, you have to use contraception" speech, but she interrupted me and said that she knew about that, and in fact, every time she does get pregnant, she just has an abortion. "In fact," she said, "I had one just last month and, now that you mention it, ever since, I've had this smelly, white discharge. Do you think it might be serious?"

That's when I realized that even this, even this simple, straightforward cardiac cath admission, was going to require a pelvic exam and that even though I was post-call and exhausted, I was going to have to stay late and do it myself. But what could I do? You can't put something like that off; if she's got PID, we'd have to cancel the cath and start her on antibiotics. And a pelvic exam isn't something you can sign out to the person on call: "Oh, yeah, the cardiac cath I admitted needs to have her pre-op bloods checked, and I think she might have PID. So would you do a pelvic exam?" I don't think that'd go over real well, although I'm sure some of my co-interns have tried to pull stuff like that.

Anyway, I did the pelvic and it turned out she didn't have PID, just some nonspecific vagitch [**internese for vaginitis, an acute inflammation of the vagina**], and that wasn't going to keep her from having her cath. But it did keep me from getting home until after seven. I missed going to my grandmother's for Tuesday night dinner. And now I'm so tired, all I can think about doing is going to bed.

You know, all these pelvic exams reminded me of something. When you're in college and medical school, you read all these books like *House of God* where the interns are spending half their lives in bed with the nurses, the social workers, and all the other females who populate the hospital. By the time I started this damn internship I knew those kind of things didn't really happen, but I did expect to continue to have at least some semblance of a sex life. But this week, spending so much time in the gynecology room doing pelvics diagnosing PID, I actually realized I have less interest in sex now than I ever remember having! It's scary. All I care about is getting to sleep. This might be the reason Carole hasn't been spending too much time with me over the past few weeks. I hope this is reversible.

Well, I'm going to sleep now. I've got to get my eight hours every other day or so.

Tuesday, December 10, 1985

I am furious! I can't believe they're doing this to me! I tried to be nice to them, to do them a favor when they needed it, and they wind up screwing me, *screwing me* in return! I just might kill the chief residents, all four of them. I know I've said that before, but this time I'm serious! I think I'll spend the next few hours figuring out exactly how I'm going to do it. Yeah, that might calm me down a little. I just can't believe they're really doing this to me!

It all started this afternoon. I was sitting around the nurses' station on Adolescent, saying how I couldn't wait for the month to be over because next month I'm scheduled to be on Children's and I love Children's. And Arlene, one of the chief residents, said, "Oh, haven't you heard? We had to pull you from Children's next month and put you on 6A." She said that, after all, I had already done July on Children's and that there were some interns who were never scheduled to work there and it wouldn't be fair if I wound up doing two months and other people wound up doing none. I very calmly reminded her that I had only done a month on Children's because the chief residents needed someone to fill in when one of the subinterns didn't show up and they had promised that if I did them the favor, they wouldn't pull me from my regularly scheduled month. Arlene said she hadn't heard anything about that; all she knew was, I would have to spend the month on 6A. I should have pulled her head off right there while I had the chance.

Not only are they going to do this to me, not only are they going to take me away from the best rotation in the system, but also they're going to deprive me of working with Amy Sorenson; Amy Sorenson, who, in addition to being one of the smartest and friendliest of all the residents, also happens to be one of the best-looking. I've waited all year to work with her; I've even dreamed about it. It's one of the only things that's been keeping me going. And now they're making me switch to 6A, where my resident'll be Attila the Hun. And not only are they taking me from the best ward with the best resident and putting me on the worst ward with the worst resident, but also *they weren't even going to tell me about it until I showed up at the start of January rotation*! So what this means is I'm going to wind up doing a total of five weeks on 6A and three weeks of Children's. I'm getting screwed, and I'll tell you one thing: Even if I don't wind up killing the chiefs, you can bet I'll never be caught dead doing a favor for them again!

Friday, December 13, 1985

I'm waiting for Carole to come by. We're going out for dinner with Bob Marion and his wife, so I've got to make this quick. It looks like I won my battle, and I didn't even have to use force. When I showed up for work on Wednesday ready to go up to the chief residents' office to reenact some of the more gruesome parts of *The Texas Chain Saw Massacre*, Arlene came up to me in the hall and said, "Oh, we made a mistake. You don't really have to work on 6A next month after all." Apparently the chiefs talked it over and decided they really couldn't screw me like that. That's kind of nice, but it threw me off guard. I mean, it's completely against their nature to be nice to interns. I have to assume they're setting me up for something. So now my problem is, I'm sure they're going to try to ambush me, but I'm not sure when. It would have been easier if they had forced me to do the month on 6A and I had just killed them; it would've taken all the guess work out of the next few weeks.

Well, the bell just rang. I guess that's Carole.

It's still Friday night, or maybe it started being Saturday morning already. Yeah, it's twelve-thirty. I just got home. Carole and I had a big fight. What else is new? I'm such a wonderful conversationalist, so nice to be around. All I ever do is complain about work, and all she ever does is complain about me. Tonight she started in on me about my apartment. We went to this seafood place on City Island and she asked Bob and his wife if they knew of a place where I could move that didn't have roaches. I don't want to move! I like my roaches; they give me someone to talk to. So I said, "I don't want to move!" and that started it off. She said she understood I was working hard but that she was a person, too, and if I wanted her to be part of my life, I was going to have to make some time for her in my busy schedule. I said that life is hard enough right now for me without anybody making demands on my time. Bob's wife looked kind of uncomfortable through this, but Bob was eating it up. He wants to make this year into a book, so he was salivating with all this intimate social stuff. We fought for a while but then we made up around dessert. So now we're friends again.

Carole's given up on getting married, at least for right now. I don't know what's going to become of this, but I do know one thing: If this relationship can make it through this year, it can make it through anything.

Wednesday, December 18, 1985

Last night something really funny happened. It almost made it worthwhile being on call. Wow, what a weird thing to say. Nothing could ever make being on call worthwhile. Anyway, here's what happened: I admitted this fifteen-year-old girl who had been found unconscious in the street. Someone called EMS [**Emergency Medical Service**] and they rushed her to West Bronx. It was quickly figured out that she had overdosed on a combination of crack and heroin. They worked on her for a while in the ER and got her stabilized, then decided to admit her to Adolescent. So I went down to the ER to pick her up. That's when she woke up. Lucky me.

She wasn't what I'd call the friendliest patient I'd ever seen. What she was was abusive. She cursed out anyone who came within ten feet of her, me, the nurses, my medical student, everybody. I couldn't examine her, I couldn't even get close enough to get her vital signs. So I called the senior and she came, and after she got cursed and threatened for a while, she said, "No way! I'm not touching her!" So here it was, nearly two in the morning, and we've got this lovely young woman in our treatment room who isn't exactly happy to be there, and we're supposed to do something to make her better. So the resident said, "If we can't examine her, we have to call the person who has ultimate authority. Who's that?"

I said it was the attending. She said, "Well, call him and tell him to come in and see if he can talk to her." And then she left. Very helpful.

Well, I called the operator and got Hal Loomis's home number. I felt a little funny calling him that late for something like this, but when I told him the story, he said, "No problem; I'll be there in a half hour." He actually seemed happy I had called him at home and woken him up! These attendings are really weird.

About an hour later, Hal walked out of the elevator and onto the ward. I led him into the treatment room and pointed to the stretcher. She seemed to be asleep, but when we got close to her, she opened her eyes and started yelling at us. "You stay where you are, you fucking son of a bitch. Don't come any closer or I'll kick you in the balls, you asshole." And, without missing a beat, Hal yelled back at her, "You watch what you say, you little bitch! We don't want to have anything to do with you either, but you're here and you're sick and it's our job. So you better let us do what we have to do or we're going to tie you down and do it anyway. Now, which way do you want it?"

Well, it was great logic, and I think it would have worked if she'd been in her right mind, but, of course, she wasn't. So when Hal and I got close to her again, she started punching and kicking and biting us. But with him there holding her down, I at least could get some sort of a physical exam done. After about a half hour of that, I said I was finished, and we tied her down to a bed in one of the rooms with leather restraints. Then Hal headed back to Westchester, I got a couple of hours of sleep, and this morning she was a real pussy-cat, believe me. She's just fine today.

So I didn't mind being on call last night. If I could watch an attending make a fool of himself every night I'm on call, I think I wouldn't even mind being an intern.

One more week to go on this damn ward. One more week and the year's half over. I can't wait!

Bob

This month marks the halfway point in the interns' trip through this horrible year. The first half of internship officially ends on December 28, and I'm sure that on that night, as they lie in their beds trying to fall asleep, or sit in nurses' stations around the Bronx trying to finish yet another admission note, many of our interns will take a minute or two to reflect on what's happened to them since July 1 and try to work up enough enthusiasm to propel them through next June.

The pediatric department's annual holiday party was held on Wednesday night, December 18. Most of the interns managed to turn out. That's surprising when you consider that a third of them were supposed to be on call and another third were postcall and many of these guys probably hadn't slept in a couple of days. So the fact that so many were there was nice.

Just watching them, it was clear that something had changed since the last time this group had met at a prearranged site for a party. At the first orientation party, the interns had seemed isolated, nervous, and scared to death.

Now, six months later, they were a strong, unified group. Tight bonds had formed among them from the mutual sharing of the good times and bad that have occurred over the past six months. On the dance floor, at the bar, sitting at the tables, there was a lot of backslapping, a lot of laughing, and a lot of inside jokes. These guys have built a strong support network for themselves; they're there to help each other out. That's exactly the way it should be at this point in the year.

Thinking back on it, these people have definitely changed. They've gone from being frightened, untrained, technically awkward but very concerned medical students to competent, overworked, and chronically overtired interns. It no longer takes them all night to start an IV or all morning to draw the routine bloods on their patients. They've become masters of scut; they've learned how to manage their time so that they no longer have to stay until eight or nine o'clock on the nights when they're not on call, as they did when the year first began. They've learned the shortcuts that are necessary to survive.

They're also beginning to feel comfortable being around critically ill patients. They no longer feel the impulse to run away as fast as they can when they hear that a three-year-old who's in the midst of a convulsion or a six-month-old with signs and symptoms of meningitis has appeared in the emergency room. They've started to be able to formulate a plan of management by themselves, not relying as much on the residents or attendings to tell them what to do and when to do it. And they're beginning to develop good instincts; they're now able to figure out which patient is truly critically ill and in need of immediate attention, and which patient is not so sick and can wait. But these skills are still in an embryonic state. It'll take a few more months before any of the interns feel confident enough to reject advice given by an attending physician. But one day that will happen. They'll suddenly realize they can do it all themselves.

That's how it was for me. I remember the night everything seemed to come together. It was the middle of March and I was working on the general pediatrics ward: the worst night of my internship. Starting in the afternoon, I had admitted patient after patient, each sicker than the last. By the next morning I had trouble remembering them all; there had been at least eight of them, with three sick enough to quality for admission to our hospital's ICU.

It was at about five-thirty in the morning when it suddenly hit me. The sun was coming up and I was finishing with my third ICU admission, a fourteen-year-old girl who was comatose and near death due to acute inflammation of her brain. She had been sick with chicken pox the week before and had now developed post-varicella encephalitis, a very rare, devastating, and often lethal complication. I had admitted her and done the entire workup by myself, including putting in an IV, drawing the bloodwork that I thought needed to be done, and performing a spinal tap. I had decided on a plan of management and had confirmed that plan with all the appropriate consulting services. And as I sat to do my admission history and physical, with the girl's vital signs finally stable, after this long and terrible night, I realized all of the sudden that I could actually do this stuff. I could be left without someone looking over my shoulder and the job would get done. And once I came to this conclusion, I knew for the first time all year that I would survive my internship.

But it wasn't until March that I reached this conclusion.

It's only December now and, although Mark, Andy, and Amy have come a long way, they still have a long way to go.

Mark came to the Christmas party with Carole. They seemed to have fun, but Carole has to have a tough time at events like this: She has to feel like something of an outsider, not being involved in medicine and knowing few of the people. And Mark has to feel a little uneasy, trying to share the experience with his intern pals while at the same time making sure that Carole is enjoying herself. They spent most of the night off to the side by themselves.

Andy didn't show up at the party until after nine. He had gotten out of the hospital late after a busy night on call, and he had stopped at home to take a shower and change his clothes before coming over. He was wearing a bolo string tie, had his hair slicked back, and was wearing a pair of horn-rimmed glasses that he hadn't worn since sixth grade. The effect of all of this was that he looked as if he were on his way to a costume party.

Andy immediately joined in with a group of eight other interns who stayed together through the rest of the evening. This group is composed of the interns who had either started the year alone, without "significant others," or, like Andy, with significant others who lived outside the New York area. These people have supported each other through the first half of the year, and they have formed very tight, close friendships.

The interns in this group have little to worry about. They may not each be feeling great right now, or be extremely happy about the prospects for the rest of the year, but they know they've got each other and they know that no matter what happens, the others will be there to help them through any bad times.

At the Christmas party, the house officers traditionally put on a skit. This year, the senior residents presented a little play about what life must have been like in the Jonas Bronck ER back in the "Days of the Giants," the phrase facetiously used to describe the times when the current attendings were doing their training. The myth about the "Days of the Giants" goes something like this: "Back when we were interns, we worked much harder than they do today. We were on call every other night, and we loved it. And when a tough case was admitted, we fought to be able to take care of that patient. We wanted to impress our chief with how good we were."

In the skit, senior residents were Alan Cozza, Mike Miller, Alan Morris, and Peter Anderson. They ran around a pretend emergency room trying to prove how *macho* each was. They got into arguments and ultimately fistfights about who would admit the critically ill patient (played by another senior resident) who was brought in by ambulance.

But the residents also went on to depict what actually occurred once those Giants got those really tough cases: They didn't know what the hell to do with them. Because the reality of the situation is that back in the "Days of the Giants," there wasn't a tenth of the technological advances that are commonplace today. In fact, pretty much all the Giants could really do was fight over the patients; there was very little that could be done to cure many of the problems presented. The skit ended with a very bitter and melancholy song about the life of the residents.

The other attendings and I all left the party early; that's also become traditional. The latter part of the Christmas party belongs to the house staff, a time for them to let loose without having to worry about being judged by their bosses standing off in the corners. The morning after, there were a lot of exhausted but happy interns running around the Bronx. They've got six months to go. In many ways, these last six months are much tougher than the first six.

Andy

Sunday, January 19, 1986, 1:00 A.M.

I started my vacation, as planned, in Portland [**Maine**] with Karen and her family. We were there for Christmas. I ate like it was going out of style, I vegged out and slept a lot, and I got to know Karen's family a little better. Three days never went so fast.

After leaving Portland, we went back to Boston but we only stayed overnight. We had originally planned to go to New Orleans, but we went to California instead. We decided not to go to New Orleans because we saw in the newspaper that it was forty-five degrees and rainy down there and we heard that one of the big college football teams was going to be in town for a bowl game and there were going to be millions of crazed football fans running all over the place. So we spent a week out in Santa Barbara instead. We stayed at Karen's sister Kathy's house. Kathy was still out in Portland with Karen's parents. My brother and his wife, Debbie, and

Karen and I shared this little bungalow with a porch in the backyard where you could sit and look out and see the Pacific Ocean in the distance. It was very quiet, very beautiful, and warm. We did a lot of walking that week; we walked on the beach and in the hills and around town. It was really a good kind of meditative thing to be doing. I had a chance to look back and think about what had happened to me over the past six months, what this internship had done to me. We watched a million movies on Kathy's VCR, just one after the next. I slept a lot, and that was very good, too, just having the chance to catch up on some of the sleep I've missed. And I balanced my checkbook, which I hadn't done in six months. I brought all the stuff out with me because I knew I wouldn't do it otherwise. And I felt like my life was a little more back in order again.

At the end of the week, we were all very sad to go home. Karen and I were still enjoying each other's company a lot. We went back to Boston, where it was frigid and bitter cold. I had a few more days there. I saw a couple of old friends, and then Karen and I packed up all our stuff and got ready to come back to New York. Karen has come out to stay for two whole months. She's doing a subinternship in psychiatry in Westchester.

I came back from vacation relaxed and happy, and I was hoping my mellowness would carry me along for a couple of weeks, at least into February, when I'm scheduled to be in the ICU. The depressing thing is that the pace of being back in the ER, the aggravations of being an intern, the frustrations that come with taking care of patients all mounted very rapidly, and it took only a couple of days before I felt like I'd never left. And it's kind of a drag. I mean, here I am, only back for a week and a half, and already I'm feeling aggravated.

Most of the patients I've been seeing have been really abnormal children, really abnormal! During OPD, I spend two out of five weekdays in clinic, and that's what's killing me! All the kids I follow now seem to be abnormal; I've picked up tons of patients who've been discarded by other doctors. I've got kids with MR-CP [**mental retardation, cerebral palsy**], kids with seizures, kids with weird syndromes, psychotic adolescents I picked up while on the ward; you name it, I've got one of them in my clinic. I seem to have no straightforward, healthy children at all.

And in the ER, well, we do see relatively normal kids there, but it's such a bad situation. The parents are exhausted, they're frustrated, they've had to wait no less than forty-five minutes before they're seen; most of the time they have to wait a couple of hours. Half of your interactions with parents in the ER are not very good. I try so hard to make things go off well, but it's so hard. By the time they get to see you, the parents are so aggravated that you get aggravated. It's just a vicious cycle.

There were a couple of bad things in the ER today. I had one kid who came in and got worse right in front of my eyes. We wound up nearly coding him. And then we had a kid with 20 percent second-degree burns to the perineum [**the diaper region**] that didn't look very nice. How do you think those got there? It was another abuse case, of course.

Then a thirteen-year-old stab victim came in. The stories are always the same with stab victims: They say they were just going to the store to get their grandmother some ice cream or something like that when somebody out of the clear blue came up to them and stuck a knife into their chest; they're always innocent. This kid wasn't really that bad. And he was about the worst we had today. I didn't have to do any pelvics. So that made it a pretty good day.

I used to get upset about doing pelvics, but I really don't care about them that much anymore. They really aren't so bad as long as you've got a kid who isn't going to be hysterical. That's about one out of every five kids. I'm not wild about doing the other hysterical four, but one of those will be only semihysterical, and only one of the other three will be completely off the wall. But you really can't blame them; most of them are twelve years old and they've never had a pelvic before, and then they find out they're pregnant. Uhh, God forbid! Anyway, it happens all the time. And sexual abuse, you know—what can I say?

We had this attending on today who was driving me up a wall! She was so indecisive, I wish I'd never asked her anything! I think she made more trouble for me than anything else. But I kind of liked her, she was really very nice, and she actually gave me a little off-the-cuff talk on pharyngitis that was very good. But every other time I asked her for help, she just wound up making everything very confusing.

I'm getting to the point where I don't want to bother with the attending, I just want to ask other residents for advice. The attendings usually wind up mucking you up, unless they're really good, and that isn't too often. I'm realizing that it's best just to listen to their advice as a suggestion and then do whatever you want to do. Shit, it's my name that goes on the bottom of the ER sheet, not theirs! [**Although the attending is supposed to be supervising the care of all patients in the ER, the house officer is the one who signs the chart at the completion of the patient's care**.] I'm the one who's really responsible!

I really can't complain about anything tonight. First, I got home at a great hour. I mean, I left that emergency room at twelve-thirty. That's almost unbelievable! And I have the next two days off because Monday is Martin Luther King's birthday and all the clinics are closed. Hallelujah! What will I do with myself with all this spare time? Sleep, probably.

Sunday, January 19, 1986, 11:30 A.M.

I was just lying in bed here thinking about how no one tells you, really, how to be an intern. They tell you what to do, when to do it, how much to do it with, and how you're not doing it fast enough, but no one tells you really how to be an intern. For instance, where do you draw the line between your own decisions and those of your superiors? Over the past few months I've come to feel comfortable with making decisions; I can deal with a lot of issues on my own now. But when the attending tells you to do something and you don't think it's exactly the right thing to do, what are you supposed to do? After all, it's your name that's on the paper, not the attending's. A lot of times it seems like the attendings don't really fully understand the case, and they make snap decisions with only a half or a three-quarter understanding, and you're the one who's supposed to carry out their orders. So what it all comes down to is, you have to decide for yourself. You're not a medical student anymore, you're really a doctor, even though you barely know how to function as one. That's what no one can tell you. It's something I can barely tell you myself.

The other night I examined a little three-year-old girl who came in with a vaginal discharge. The history wasn't suspicious at all, and there were only a couple of very, very subtle things on the physical exam aside from the vaginal discharge, but the first thing you're supposed to think of in a case like that is sexual abuse. And that's exactly what went through my mind.

But I found myself getting talked out of reporting the case to the BCW because it wasn't all that clear-cut. The attending argued that the discharge could have been caused by something other than sexual abuse. I had to agree. And I felt really pressured by the attending and the social worker, people who have had years and years more experience than I, just to let it pass, to sign it out as nonspecific vaginitis rather than sexual abuse. We talked about it for a long time, and they told me to think about what reporting it would do to the family; the child would be removed and placed in a foster home. The parents would be labeled as criminals, whether anything really happened or not. It might be years before these people's lives would return to normal. And with all that pressure, I decided to go along with them.

Now I'm regretting it. I've been thinking about this kid ever since. The attending told me to be sure to follow the girl carefully. But let's say this was a case of abuse: What if they don't come back for their follow-up appointment? What can I do then? And, of course, I called the bacteriology lab at Jonas Bronck today and was told that they have no record of receiving the GC **[GC: gonococcus, the bacteria that cause gonorrhea]** cultures. Great! I'll keep looking for them; I'm sure they'll turn up sooner or later. I hope to God the messenger didn't throw them down the elevator shaft or something like that. But if they don't show up and we never find the cultures, what can I do? And what'll happen if this girl comes back dead next week because whoever molested her decides to whack her over the head with a hammer? It'll be my fault, because I listened to the attending and the social worker rather than doing what I thought was the right thing.

This feeling I have, that I have to start making up my own mind and not relying on other people, it's really something that can't be taught beforehand. I'm just realizing it myself, and I've been doing this for six months now.

I got on the elevator in the DTC building [**the clinic building at Mount Scopus**] the other day and the elevator stopped at one of the adult floors. This middle-aged man came on with these two middle-aged women, and he said something like, "All they see when they look at you is dollar signs." Then he said, "Look at their mistakes. They fill the graveyard. They don't give a damn. All they care about is money. I don't trust doctors one damn bit anymore." This guy was standing three inches from me! And I was really biting my tongue. I felt like saying, "Look, there are some bad apples out there who suck, who are only in it for money, and who don't give a shit about people. But most of us really do care about our patients."

I don't know, I find myself feeling defensive about organized medicine and at the same time being more disillusioned about it than ever. I do look at the mistakes that are made and the horrible outcomes that result, and yes, our mistakes do end up in the graveyard, but they dot them, they don't fill them.

Well, I've gotten a little off the subject. Anyway, I don't think there's any way that people can be taught about what it's like to be in the uncomfortable position of having to start to use their own mind but having very little to base decisions on. There's just no way anybody could have prepared us for this transition from the little puppy dogs who do everything the attending tells us to independent doctors who wind up being very uncomfortable with some of the decisions we have to make. I'm constantly feeling as if I've got a green belt in karate, that I know enough to kick someone but I might break my own foot doing it.

Amy

Friday, December 27, 1985

It snowed yesterday for the first time this winter. Sarah's amazed. We took her out in her stroller a little while ago and she kept looking down at the ground and looking up at Larry and me as if to say, "Where did all the grass go?" It's funny to watch.

I've been in the nursery at Jonas Bronck for a few days now. So far it's been a mixed experience. I'm assigned to the well-baby nursery, which is nice. I spend most of my time examining newborns and talking to their mothers. That's what I really liked about pediatrics in the first place, and it's nice to have the chance to do it without all the other non-sense that usually takes up our days. So that part of it is good. What I don't like, though, is that my night call is in the neonatal ICU. It's frightening in there! And it's harder for me than for the other interns because, since I'm only in the NICU at night, I don't know the sick ones very well. All I

know about them is what the interns sign out to me, and it's
impossible to get a really complete sign-out on a patient who
has a hundred different problems. So that's frightening to
me, but what can I do? It could be worse; I could have to
spend all my time in the ICU.

Another bad part about working in the well-baby nursery
is that we're always on call to the delivery room. If there's a
premature baby being born or a baby who's in distress, the
resident and I get called to come to the delivery. It's not
really that bad, though, because during the day there's usually a fellow [neonatal fellow, a physician who has completed a pediatric residency and is getting specialty
training in neonatology] or an attending around, and one
of them usually comes in with us. If they weren't there, it
would be terrifying!

I have to admit, I've been lucky with my schedule over
the past couple of months. I've worked with very good residents and I haven't been on the hard wards or had a lot of
bad patients. I guess I should say that I finished on Children's
last week and Angela [the young girl with neurofibro-
matosis] was still alive. I heard she got worse the day after I
left, though. She had another very long seizure and they had
trouble stopping it, so they transferred her up to the ICU.
They had to anesthetize her to get the seizure to stop [gen-
eral anesthesia is used as a last-ditch effort to stop
intractable seizures only after every other treatment
modality has failed]. The intern who picked up my patients
told me they don't expect her to survive much longer, only
another few days at most. It's really sad; one month ago, she
was a completely normal child. Now she's almost dead.
That's not supposed to happen to children.

I was on call Wednesday, Christmas Day. It wasn't so bad,

since we don't celebrate Christmas, but it was like working an extra weekend day. The hospital was completely dead, even deader than most Sundays. But babies don't know anything about holidays; they crump whenever they feel like it. I did get a couple of hours of sleep that night and I guess I should be thankful for that, but I can see that night call during this month is going to be terrible.

I had only one admission to the unit on Wednesday, a thirty-weeker [**thirty-weeker: a baby born ten weeks prematurely**] who did pretty well. We were in the DR when he was born. The obstetric residents thought he was only going to be about twenty-six or twenty-seven weeks; my knees were shaking while I stood in the delivery room waiting for him to come out. The resident and I were very relieved when we saw such a big baby come out. He weighed about thirteen hundred grams, which is gigantic for the NICU. And he didn't get too sick: He had a little bit of respiratory distress but nothing terrible. All he needed was a little extra oxygen, so we put him in a headbox [**a cylindrical Lucite box that covers the head of an infant and through which oxygen can be provided**] with 40 percent oxygen. [**Normal room air contains 21 percent oxygen; therefore 40 percent oxygen provides about twice the normal concentration of oxygen.**] He never retained CO_2 [**babies with respiratory distress syndrome, a major complication of prematurity caused by underdevelopment of the lungs, develop a buildup of carbon dioxide, or CO_2, in the blood**], so we didn't have to intubate him. He should do fine. His mother is seventeen years old and already has a one-and-a-half-year-old at home. She lives with her mother, who essentially takes care of her and the baby as if they were siblings. It's a funny social system here in the Bronx. Most of our mothers are under twenty, and most live with their mothers, who wind up taking care of the children.

Well, the apartment is quiet. Sarah's asleep, Larry's watching TV in the living room. I'm going to go to sleep. I've got to be up early tomorrow morning so I can be on call.

Saturday, January 18, 1986, 10:00 P.M.

I haven't recorded anything in a while. I've been very tired and very busy. I'm really enjoying working in the well-baby nursery; it's the first thing this whole year I could actually see myself doing for the rest of my life. The problem is, there's no way to do it without doing a fellowship in neonatology first, and that is something I definitely do not want to do. So once again, I'm kind of stuck.

I've gotten along very well with a lot of the mothers. They seem to trust me. They trust me even more when I tell them I've got a baby of my own who's almost nine months old. I guess they feel they can identify with me. Frankly, I'm not sure how you can be a pediatrician and give advice to mothers without having your own child. Anyway, it's been a very rewarding experience.

I've pretty much gotten my work down to a routine. When I arrive in the morning, I look at the list of babies who were born the night before. All of these kids need to have physsies **[physical exams; all babies get examined within twelve hours of delivery and then again right before discharge]**. I find the babies and do the exams. When I'm done with those, I find the list of babies who are supposed to be discharged that day. I find those babies, and one at a time take them out to their mother's bedside and examine them right in front of their mothers. I found that that gives the

mothers the chance to ask about anything they don't understand or anything they're concerned about.

It's amazing how many strange things these women
come out with. I had this one woman, a nineteen-year-old
who had had her first baby, who asked me about the strawberry hemangioma on her baby's back. [**Strawberry hemangioma is a birthmark composed of a mass of blood
vessels; they are very common and usually are of no medical significance; most disappear by the time the child is
six years old.**] I told the mother it was just a birthmark and
that it wasn't anything to worry about. She asked me three
times if I was sure that that was all it was, and each time I told
her I was positive. Finally, I asked her why she was so worried
about it. She wouldn't say anything at first, but finally I got
her to tell me the story. She said she had heard that some
people with AIDS had a skin disease that might be the first
thing that's noticed. I told her that was true, that the skin disease was called Kaposi's sarcoma. I also told her that babies
almost never got Kaposi's and then I asked why she was so
concerned. At first she said it was because the baby's father
had been using drugs for years and she was worried that he
might have AIDS, that he might have passed it along to her,
and that she might have passed it along to the baby. I worked
on her for a while, and I finally got her to admit that she had
used drugs a few times about a year ago and that she and the
baby's father had sometimes shared needles. Ever since, all
through this pregnancy, she had been scared to death that she
had AIDS.

I spent over an hour with her. I asked about all the signs
and symptoms that might indicate AIDS. She didn't have any
of them, and I told her that was a very good sign. But she

said she had been having trouble sleeping at night for a few months because she was so worried and that it was starting to affect her schoolwork. She goes to Bronx Community College. She told me she wanted to be a lawyer but she honestly didn't think she was ever going to make it because she was going to die of AIDS. And then, when the baby was born with the strawberry hemangioma, she had become convinced that not only was she going to die of AIDS, so was her baby. She started crying and I held her hand and comforted her.

I waited until her crying stopped and then I told her that if she wanted, I could take some blood to see if she had antibodies to HIV. She told me she'd thought a lot about getting tested but she was afraid to. She said she didn't know what she'd do if she turned out to be positive. I told her that was a problem, but I pointed out that she was already suffering and it might all be for nothing; there was a good chance, after all, that she'd turn out to be negative. So I guess I talked her into letting me do the test. I had her sign the consent form, and then I drew her blood. I wore gloves when I was taking it. I felt funny putting on the gloves; it was as if I were saying, "I've been telling you I don't think you have it, but I'm not taking any chances." She didn't say anything about the gloves. I don't know; maybe we make too much out of feeling guilty. So far, whenever I've worn gloves, none of the patients or their parents has said a word.

Anyway, I think I did some good for that woman. Here she had been coming to obstetricians for months, always with this dread fear, and nobody had found out anything about it. And just because I spent a little extra time with her, I was able to discover that her life was being completely disrupted by something that might be totally avoidable. I haven't gotten the results of the blood test yet. But I'm going to see her and the baby in clinic sometime next week, and hopefully by then I'll have the answer. I felt really good about that one.

I've had a couple of cases that didn't turn out that well, though. And one of those made me feel as bad as that last case made me feel good. During rounds our attending, Joan Cameron, always tells us we should try to push breast feeding whenever we get the chance. I have mixed feelings about breast feeding. I mean, I know it's the best thing for the baby; it's supposed to be helpful in preventing infections and things like that, and it's also supposed to help the bonding process between mother and infant. But it's not the easiest thing to do. A woman really has to be committed to breast feeding, and she has to have a lot of support from the people around her. If she's kind of wishy-washy about it, it's just not going to work out.

Anyway, last week I was talking to this woman who asked me about breast feeding. I gave her the party line: I said yes, it's the most important thing you could do for your baby. Then she asked if I had breast-fed my baby (I had already mentioned to her about Sarah). And I had to say that I did it for a few weeks only and then stopped because I had to start my internship. And she said something like, "You doctors are all alike! You tell us to do things you wouldn't be caught dead doing yourself!" And she said some other things that weren't very nice. Basically she called me a hypocrite and she immediately asked for a bottle of formula.

I knew she was right, and she hit a nerve. I mean, I would have liked to have breast-fed Sarah for longer if I'd had the chance. It makes me pretty angry. Here we are, being told by our attendings that we should advocate breast feeding, but there's no way I would have been able to do it with my own baby. How can you breast-feed if you're on call every third night and there's no place in the hospital to keep your baby while you're working? That woman was right, it *was* hypocritical for me to suggest she do something I couldn't do, and it's very hypocritical for our faculty to try to get patients to do something that's best for their babies and not give the house staff the same opportunity. So that situation didn't work out so well. And I'm still angry about the whole thing.

My night call is just about what I expected. I've only gotten sleep a couple of times on nights I've been on call. I'm finding something out: I really need only about two hours of sleep to function well the next day. But those two hours have to be between four and six in the morning. If I'm up between four and six, I'm just about worthless the next day. If I sleep those two hours, even if I haven't seen the bed the rest of the night, I'm fine.

And doing night call in the NICU hasn't made me feel any more comfortable about working with these tiny babies. If anything, I've become more terrified. The unit is brand new; it just opened a couple of months ago, so everything is state of the art. And these babies are so sick! We've had three deaths so far this month—two preemies and one full-term kid. I was on call the night the full-term kid was born. That's something I won't forget for a long time!

We were called to the DR because of thick mec and late decels [**late decelerations: a pattern on fetal heart tracing indicating fetal distress**]. The obstetricians decided to do a stat C section and they pulled out the baby, who was covered with mec. I tried to suction her mouth while she was still on the table, but I guess I didn't get all of it out because she was in respiratory distress almost immediately. [**Actually, the baby had probably already aspirated meconium prior to delivery; in this case, suctioning of the oropharynx prob-**

ably didn't provide any help in preventing what subsequently happened.] Eric Keyes was the senior on call with me, and he was on the baby as soon as she hit the warming table [**the table in the DR on which the baby is placed following delivery**]. He intubated her and started suctioning out her airway through the ET tube. He was getting tons of thick mec out. In the meantime, I was listening to her heart. She was really bradycardic [**had a low heart rate**], so Eric told me to start a line and get ready to push meds. I hadn't ever started a UV [**umbilical vein**] line myself, so he talked me through it as he was suctioning out the trachea. When I finally managed to get something in, we changed places so Eric could push the first round of meds. The airway was pretty clear by that point, so I started bagging the baby [**pushing oxygen through the endotracheal tube, using an ambubag to generate pressure**]. The heart rate came up a little, to about 80 [**the normal heart rate for a newborn is 120 to 140 beats per minute**], and Eric decided that we'd better get the baby out of the DR and into the ICU right away, so we put the baby in the transport incubator and ran with her down the hall to the unit.

We worked on her all night. We called the neonatal fellow at home, and he came in to help. She had severe respiratory distress and PFC [**persistence of the fetal circulation**]. We were having trouble ventilating her and getting her blood circulating. We put her on a ventilator and had it turned up to very high settings. [**To ventilate a child with mec aspiration, it's often necessary to use a great deal of pressure with which to push oxygen into the lungs. Meconium causes the lungs to become very stiff, and the pressure is necessary to get them to expand.**]

At about four in the morning, she crumped. Eric decided

she had a pneumothorax [**a collapsed lung, caused, most likely, by the high ventilator pressure that was being used to force air into the baby's lungs**], so he put a chest tube in and she immediately looked better. But she was still hypoxic all night, and at about five in the morning Eric and the neonatal fellow decided to start her on tolazoline, which is supposed to help PFC. It didn't do her much good. She crumped again at about eight, just when the day crew started showing up. It was another pneumothorax. I didn't stay any longer than that. I had to get out to the well-baby nursery and start doing my physsies. The baby died a little later that morning.

It was terrible. She had been completely normal. If she hadn't gotten all fouled up with meconium, she probably would have been a normal child.

That baby's mother was put on the GYN ward, so I didn't get a chance to talk to her. They did that so it would be easier for her; it would have been very hard if they'd put her on the regular postpartum ward and she had to be surrounded by all the new mothers with their healthy babies. I don't know what I would have said to her if I had gone to talk with her. Nothing seems right.

I heard that Angela died a little over a week ago. She spent her last couple of weeks in the ICU at Mount Scopus, comatose. I never thought there'd be so many deaths in a pediatric internship!

Anyway, I've got to go to sleep. I'm getting tired and I'm on call again tomorrow, so I've got to get a good night's sleep. I've got another week to go in the nursery, then I have two weeks in the OPD and then vacation. I hope I make it until then.

Sunday, January 26, 1986

There are so many crack users around. There are six babies who've stayed in the nursery the entire month. The mother of each of these babies is a crackhead and the babies have been taken out of their custody by the BCW. They're all waiting for foster placements, but it's hard to find homes for these kids because there's a good chance they're infected with HIV. All of them spent their first few weeks of life withdrawing from drugs. It's sad. For a lot of these children, their lives are already over before they even had a chance to start.

The ICU has been pretty quiet over the past week. There was an outbreak of naf-resistant staph [**a type of bacteria that is insensitive to nafcillin, the antibiotic that is most effective in treating staph**], so they had to close two whole rooms [**closing the rooms and cleaning them is the only effective way to prevent sick newborns from getting infected with the bacteria**]. That cut the census by about half, so taking call in there wasn't so bad. I even got a few hours' sleep the last two nights I've been on. So all in all, it's been a pretty good month.

We're going out to New Jersey to visit my father this afternoon. Sarah was nine months old yesterday, so my father's going to have a little party for her. We haven't seen him in over a month; he's starting to feel like we're trying to avoid him. He still doesn't understand what it's like to be an intern. He thinks I've got a lot of free time and that we're just doing other things rather than coming out to see him.

Well, I'm going to stop now. Tomorrow I'll be back in the Jonas Bronck ICU. Seven months down, five more to go.

* * *

Oh, one more thing. That woman who thought her baby's strawberry hemangioma was a sign of AIDS showed up at clinic this week. I called the lab to get the results of her HIV testing, and guess what? She tested negative! When I told her, she hugged me and kissed me. That's the first time this whole year when I really thought I had done somebody some good. And it happened only because I took the time to sit and listen to what she had to say. It had nothing to do with medicine.

Mark

Thursday, December 26, 1985

I was on call Tuesday for the first time this month, and it wasn't too bad. It should have been great: It was Christmas Eve, the ward was quiet, and I had only one admission. I should have gotten six hours of sleep at least, right? Wrong! It was one of those cases where if anything can go wrong, it will go wrong. It was a two-and-a-half-year-old sickler with pneumonia. She was called up from the ER at about eleven-thirty, so I figured great, I'll go down, bring her up to the floor, do the workup, start her on some antibiotics, and be in bed by one. Of course, that's not even close to what happened. First, I got down there and found that no one had been able to get an IV into her. Everyone had tried and everyone had failed. So they figured what the hell, let old Mark take the kid up to the ward and have a crack at it. How nice of them! But it turned out okay, because you know what? I got it in on the first stick. That's right, the very first stick! I'll tell you, I'm becoming the King of Scut. It just shows you that if you take a plain, ordinary, moronic intern and make him do the same things over and over again until he loses his mind, you can teach him to do almost anything. I think that now that I've mastered IVs, I might take up neurosurgery in my spare time.

I'm getting off the track here because I'm a little tired. Anyway, so I brought her up to the floor and got the IV in and did the whole workup, and by two o'clock everything was done except a urinalysis. I spent most of the rest of the night chasing after her with a urine cup, trying to get some of her precious body fluid. Yes, my mother sure would have been proud of me!

They hadn't been able to get any urine from her in the ER. I didn't want to start her on antibiotics until we had a sample of urine because she had had a UTI [**urinary tract infection**] in the past, and if she had one again, we needed to know about it. I wasn't having trouble getting the urine because she wasn't peeing; it was that she wasn't real happy about peeing into any kind of container. Right after we got her up to the floor, she was standing on the scale in the treatment room and she let loose a stream, so I ran over with a cup just in time for her to pee all over my hands. No urine ended up in the cup, of course. Then we decided to straight-cath her [**place a sterile catheter through her urethra and into her bladder in hopes of obtaining clean urine**], but just as soon as I got close to her with the catheter, she started to pee straight up into the air. I managed to catch some of that in a cup, and I ran off to the lab to analyze it. It turned out that her urine was clean as a whistle. By that point it was about five in the morning. I got a total of two hours of sleep.

I found out later she was a patient I had taken care of in

August on Infants'. Now she's graduated to Children's. These kids keep following me all over the place. Next thing you know, Hanson'll show up again. Hanson! Now, there's someone I haven't thought about in a while! You know, no one's heard anything about him since I discharged him from Jonas Bronck in October. But I know he'll turn up again, you can be sure of it. It'll be the busiest night of the year; there'll be hundreds of admissions to take care of, and he'll come toddling in and take one look at me and crump right there and then!

I've been totally and completely terrified of Alan Morris **[the attending in charge]**. Monday, on the first attending rounds of the month, he asked me to tell him about my patients. I went to present my first kid and I started off by saying, "This is a six-month-year-old-month old . . ." I just couldn't get the words to come out right. I got so tongue-tied I finally said, "Forget it! I can't present anybody to you! You make me too damn nervous!" That was good because it loosened everybody up. Alan seems to make a lot of people uptight. I'm not sure what it is about him that does it. Maybe it's the whip he brings to rounds with him. Or the buzzards who are always circling over his head. I don't know why, but he definitely makes me uptight. He's a great teacher, though; so far, rounds have been excellent. I had to present my sickler to him this morning and I managed to get the words out, but I was still nervous. Then we wound up talking about sickle-cell disease and he happened to hit on the one area I actually knew something about. He grilled me for about a half hour and I think I did a pretty good job. In fact, he must know I need some positive reinforcement, because for the first time since I've been here, I actually heard him give someone—in this case, me—a compliment. He said something like, "I don't care what everybody else is saying about you, I think you're doing a reasonably good job." Talk about a vote of confidence! I guess it's better than having him say he thought I was a complete idiot!

The floor was a real disaster today; poor Ron was getting creamed! There were four admissions, and each one had a bizarre story. One of them was an eight-year-old with subaortic stenosis [**an obstruction to the flow of blood below the aortic valve; this obstruction prevents blood from getting from the left ventricle of the heart out to the rest of the body**] who was only mildly symptomatic but who was admitted for surgery anyway. Ron and Amy, our resident, did a complete workup, history, physical, labs, the works. When they had drawn his type and hold [**a specimen of blood to be sent to the blood bank so that blood for transfusion could be prepared**], the mother said, "What are you doing that for?" Martha told her it was for the blood bank and the mother said, "Well, you don't have to send it. Don't you know we're Jehovah's Witnesses? There's no way you're going to give my child any blood, and that's final!"

They went crazy. Ron was ready to reach into the cardiologist's mouth and tear out his vocal cords. And then this whole big thing started with the cardiothoracic surgeon, two or three anesthesiology attendings, the cardiologist, and us. The Anesthesia Department refused to do the surgery without the option of using blood if it was needed. They had to call the hospital lawyers and wait for a ruling. The whole thing took hours, and the end result was that the kid wound up going home. Amy and Ron were pissed off, the cardiologist was pissed off, the CT surgeon was pissed off, and the anesthesiology people weren't exactly happy.

I sat through all this in kind of a daze because I was so

tired. I've got to try to get more rest. Well . . . maybe next year.

Tuesday, December 31, 1985, 11:30 P.M.

I know it seems pretty strange, but here it is, eleven-thirty on New Year's Eve, and I'm lying in bed, talking into this stupid machine. I'm too tired to go out, so I'm here all alone. Carole went to a party by herself. I was supposed to go with her, but I called her a couple of hours ago and told her I was just too tired to make it. I'm pretty pathetic!

I had a long night last night. It took forever to finish my work today, and just as I was about to leave, a nurse came running out of a room yelling, "Hurry up, she's not breathing right!" So I calmly got my stethoscope and walked into Cassandra's room, and there she was, sure enough, breathing at a rate of eighty to ninety. Now even I know that eight-year-old girls aren't supposed to breathe at a rate of eighty to ninety. I wasn't sure what was happening. She's got osteogenic sarcoma [**a malignant tumor of the bone**] and she isn't expected to live very much longer, but I at least expected her to make it into 1986. When I came in and found her breathing that fast, I figured maybe she was having a pulmonary embolus [**a clot in one of the lung's major arteries**]. But she had equal breath sounds. We did a whole workup and didn't find anything. She's not my patient, but we've all gotten to know her. I just called and found out that she's still alive and she seems to be reasonably comfortable, which is reassuring.

I don't know about these terminal patients; it's really draining taking care of them. You don't even want to go near the room because you know there isn't anything you can do to help, and whenever you do go in the room, it's to do something terrible, like draw blood. It's very frustrating. The only thing we can do is try to make the last few months as comfortable as possible for her. If we can do that effectively, then we've really done our job. Dealing with these kinds of issues is really the hard part of this year.

There, now I've really cheered myself up! I'll tell you, I'm not sorry to be seeing 1985 end. In 1986 I'll be an intern for only half the year. That's not so bad. I'm going to sleep. Good night. And Happy New Year, tape recorder! Now, that's really pathetic!

Tuesday, January 7, 1986, 9:00 P.M.

This has been an interesting couple of days here. On Friday morning I had to get some blood from a patient before attending rounds started. I was late, and I was worried that Alan was going to yell at me. He still scares me to death. I was postcall and really crazed and I guess I hadn't eaten in maybe sixteen hours. So I went into the patient's room and started drawing the blood, and pretty soon I started realizing that I was feeling kind of light-headed. Really light-headed! So light-headed, in fact, that I grabbed on to the patient's mother, who was helping me hold the kid down. She, of course, thought I was coming on to her, but I reassured her that I wasn't trying to do anything nasty, I was merely trying to prevent myself from collapsing in a heap on the floor. I told her I'd be fine just as soon as I finished drawing the blood. I'm sure that reaffirmed her faith in me as her child's physician!

Anyway, I drew the blood, got it into the tube (which I consider quite a save, considering how hard a stick this kid was), and then I started to kind of sort of lie there on the floor feeling very dizzy. Everyone came running; I thought they were going to call a code and start full-scale resuscitation on me. But they didn't; I guess they realized that I hadn't arrested, I had just fallen over, so they got me into a wheelchair, and the nurses checked my blood pressure, which was normal, and then they put me in the house staff room, where I collapsed on the couch. I felt dizzy every time I tried to lift my head. This was, needless to say, somewhat anxiety-provoking.

I stayed in there for a few minutes and then I tried to get up so I could get to attending rounds. I got myself in a sitting position and started to cave in again when Alan Morris showed up. I said to him, "You know, I'm really not feeling too well." And he looked at me with a very serious face and asked what was wrong. I said, "Well, I've been dizzy for the past fifteen minutes." Then he said (in a formal-sounding voice), "I suggest you continue to rest; if this persists for a few more minutes, I recommend that you be brought down to the emergency room for evaluation." All in his usual righteous tone of voice.

Well, he went out and came back a minute later with the wheelchair and said (in the same formal-sounding voice), "You know what? On second thought, it's been long enough. If you had been out in the street somewhere or in any place other than a hospital, you would have already been brought here by ambulance. I think we'll take you down to the ER," which he did. By the time I got down there, I was completely white and really uncomfortable. They slapped me down on one of the stretchers, they stuck an IV in me (at least I didn't have to do that one myself), they drew blood; I got examined by one of the ER attendings, I got examined by one of the residents in neurology, and God help any of us if we really have a neurological problem and we have to be taken care of by a member of the neurology house staff, because this guy turned out to be pretty hopeless.

Well, hours and hours passed; many people came and went (actually, it was nice to get all that sympathy). My student, who generally is a moron, stayed with me the whole day and was very sympathetic. And finally, at the end of the day one of the adult neurology attendings came by. I told him what had happened, he did a quick exam, found everything to be normal, and told me I'd hyperventilated and that I should stop consuming anything with caffeine in it.

Hyperventilation—how about that? He probably was right, too. I was pretty crazed when I was drawing that blood, I was worried about what would happen if I made it to attending rounds late. But can you imagine an intern having to eliminate caffeine from his diet? Caffeine is the only thing that's been keeping me alive. If I stop drinking coffee, I'll just lapse into a coma and never wake up.

Anyway, so I left the ER at about six and Carole, who was coming to see me for the night, picked me up and carted me home. Ever since, I've been getting total abuse from my "friends." That was my emergency room event for the year. It was actually really interesting being on the other side of the bed. As a patient, you realize how sensitive you are to what everyone is saying and how they're saying it. Really, it was a very interesting experience to have had. And I hope I don't have it too many more times.

I was on call Sunday. It was a pretty good day, only two admissions. Then yesterday, I got home after being on call and I found out my phone wasn't working. It turns out it was some problem with the phone company's cable, but I didn't know that; I thought it was something wrong with my phone. I called them and made an appointment for someone to come to my apartment today at four-thirty. So I was in a rush all day, I knew I had to get out of the hospital early. And, of course, when you're in a rush, everything goes wrong: I went out to my car this morning and found that the damn thing wouldn't turn over. The battery was dead, I'm sitting out there fuming, it's freezing cold, and the damn battery's dead! I called the AAA and wound up getting to work forty-five minutes late. All because I was in a hurry.

And then attending rounds went on forever! We would have gotten done early because there had been only one admission last night, but then my moronic medical student asked Alan for a lecture on static and cidal levels of antibiotics [cidal: the concentration of antibiotic needed to kill bacteria; static: the concentration needed to prevent the bacteria from reproducing], something I'd heard eight times already. And, of course, I couldn't get up and leave. No, I had to sit there or face the Wrath of Alan. It took a whole two hours before we were done.

Then I called the telephone company to tell them I was going to be late, and could they come later, and they told me "Oh, didn't you hear? The problem was on our side, not yours. Your phone works perfectly now!" So I built up all that serious aggravation for nothing!

So, to review the past week, I've had one episode of hyperventilation, which earned me a trip to the ER; one episode of phone failure, which nearly earned me a nervous breakdown; and one episode of battery failure, which I've taken care of by getting a new battery. And who's to say what lies ahead over the next couple of days; or months; or years, for that matter, in this exciting borough I hope to get out of sometime soon, or at least before I go completely crazy! I'm telling you, one day I'm just going to get this enormous ulcer and bleed right out on the floor! It's only a matter of time!

Wednesday, January 22, 1986

Fine. Now I've developed an allergy to something. My nose has been running and my eyes have been itching like hell all night long. I may not survive this month. If I were an insurance company, I certainly wouldn't allow me to take out a policy!

I guess you can say yesterday was just another typical day. I admitted this kid who was dehydrated. They thought he had mononucleosis in the ER, but when his CBC came back, it showed pancytopenia [lack of all types of blood cells] and lymphoblasts [immature white blood cells]. A diagnosis of leukemia was entertained (what an ironic expression). And unfortunately, this morning the diagnosis was confirmed. Of course, the family's really nice.

Sunday, January 26, 1986

So tomorrow I start my month in the NICU. I'm ready to jump out the window. It won't be too bad, though. I have clinic tomorrow afternoon, so hopefully I won't be around long enough to get *really* nauseated. [Obviously, an afternoon in clinic exempts the intern from the hospital for the time he or she is scheduled to see outpatients in the clinic.]

I guess I'm going to miss Children's. It really is a good place to work. Alan Morris was an excellent attending. Even though he still scares me to death, I really learned a lot from him this month. When I went up to him on Saturday to thank him for the month, he told me I should have more confidence in myself because he thought I was a good intern. I think I've figured out what it is about him that makes me nervous: He reminds me of my father, whom I was always terrified of. I was always afraid he was going to yell at me for not eating my vegetables or something.

The main reason I liked Children's this month, though, was because of the resident. The resident was wonderful. She's just totally wonderful in all ways. She's smart and she's an excellent teacher. She's calm and she doesn't get upset no matter what happens or how many mistakes I make. And she's great-looking, too, which certainly helps. It's just too bad I won't get the chance to work with her again this year. And I swear to God, if she weren't married . . .

Bob

We're getting into the seriously depressing part of the year now. New Year's Day marks the beginning of Intern Suicide season, the time when we really have to start worrying about the house staff's mental health. There are a number of reasons why January and February are so bad. First, exhaustion is cumulative, and the interns have now built up a six-month supply. They're chronically overtired and can't get themselves too enthusiastic about anything. This exhaustion affects all aspects of their lives: They don't have energy to socialize, so they completely lose contact with family and friends; they eat too much junk food and get little or no exercise, so they wind up gaining a ton of weight. This causes them to feel down about themselves and to lose confidence.

Second, although everybody around them is celebrating the end of one calendar year and the beginning of the next, the end of the internship year hasn't even yet appeared on the horizon. There's just about nothing for these guys to look forward to right now other than another half year of the same shit over and over again. So they develop a feeling of desperation, and that feeling is compounded by the fact that they know there's nothing they can do to make the time move any faster.

Finally, the environment seems to be conspiring against them. The weather this time of year is horrible. It's constantly freezing cold, and the city is frequently getting pelted with snowstorms. It gets dark so early that the house officers can go for weeks without ever actually seeing sunshine; they get to the hospital so early that it's still dark and they come home again at night, after the sun has set. So the house officers live in a constant world of cold and darkness, and there's nothing more depressing than that.

Although January is bad, it's nothing compared with February. In January, there's still some semblance of a "spark" left within the bodies of the interns, the last vestige of the excitement that accompanied the holiday season. The department's Christmas party did the whole staff a lot of good; there were a couple of weeks during which everyone seemed a little happier and a little calmer. But it was really short-lived. And usually by February 1, any spark of excitement has been snuffed out.

A fair number of pretty strange things happened during January. The strangest involved Andy Ames, one of the interns who's in Andy Baron's circle of friends. The story started like this: At the beginning of January, Andy Ames and one of the female senior residents were working together on the Jonas Bronck wards one night and admitted a six-week-old girl with fever. Because a significant percentage of these infants will be shown to have a serious bacterial infection in their blood or spinal fluid, it is policy that all babies under two months of age who come to the emergency rooms with fever routinely get admitted to the hospital. Blood, urine, and spinal fluid cultures are taken, and the infants are started on intravenous antibiotics.

Anyway, Andy Ames and the senior resident were trying to get a sterile specimen of urine from this little girl by doing a straight catheterization, a procedure in which a plastic tube is inserted into the urethra and passed up into the bladder. The cath went pretty well, and they managed to get an adequate sample of urine for culture and urinalysis. But the mother, who was standing in the treatment room the whole time, went nuts when she realized what Andy was doing. She accused him of sexually molesting her daughter and of "ruining" her for life. The mother yelled and screamed for most of that night, becoming more and more agitated as time passed. Early the next morning, she went to Alan Cozza and the hospital administrators to complain. When the situation was assessed, it was carefully explained to the woman that what Andy had done was completely aboveboard and standard treatment and did not in any way constitute sexual molestation. The mother continued to yell that Andy had "ruined" her daughter and that no man would ever want her after what he had done. The administrators continued throughout that day and the next to try to calm her and explain the anatomy of the procedure to her. When it finally became clear to the woman that she wasn't going to get any satisfaction from the hospital employees, she decided to take matters into her own hands: She began to threaten Andy Ames with bodily harm.

From then on, things became exceedingly weird. While the baby was in the hospital, the mother told Andy every time she saw him that she was going to sneak up behind him when he wasn't expecting it and stick a butcher's knife into his back. She also told this to everyone else who was hanging around the ward, the house staff, the medical students, the nurses, even some of the other parents. Since the baby was better and no sign of bacterial infection had been found, Alan Cozza decided to discharge the child a day earlier than usual. He hoped that with the baby and her mother out of the hospital, some of the pressure would be removed from Andy, who, needless to say, was feeling quite persecuted by all this. But discharging the baby didn't help; the woman managed to find other ways to drive Andy crazy.

After discharge, the baby's mother began to call the ward asking for Dr. Ames. When Andy got on, she'd repeat the threats. She somehow got the number of the residents' room and left cryptic messages for him with Lisa, the house staff secretary. She even managed to get Andy's home telephone number and left messages on his answering machine.

At about this time, Alan Cozza, concerned about what was happening to his intern, began investigating this woman's background. Not surprisingly, he found that she had a long psychiatric history and had been diagnosed as having paranoid schizophrenia. Then, about a week after the baby was discharged, a call came for Andy in the residents' room. The person identified himself as the woman's psychiatrist. He explained that the woman had told him exactly what had happened and had laid out in explicit detail exactly what she was going to do to get back at "that intern who ruined my daughter." He told Andy that he was concerned about his well-being because she was angrier and more agitated than he'd ever before seen her.

This was all Andy needed. If he hadn't been worried about all this before, the psychiatrist's call certainly pushed him over the edge. And apparently there was very little at that point that anybody could have done. The woman wouldn't voluntarily consent to hospitalization in a psychiatric facility because she didn't think of herself as sick. Her psychiatrist, although truly concerned about Andy, was unwilling to proceed with forcing her into institutionalization against her will. He said he simply hadn't accumulated enough evidence yet to justify such a move. And so, during January, after work every day, one of the other members of the house staff had to walk Andy out to his car in the parking lot. The interns took turns staying over at his apartment. He had his phone number changed and made sure the new one wasn't listed. And all of this certainly took its toll on him. He began looking terrible: He was already exhausted from the usual intern routine, and he barely had enough strength to get through a typical day. But now he was no longer able to sleep even on the nights when he wasn't on call because he was so worried.

The story finally came to a head in early February. The baby's mother showed up in the residents' room at Jonas Bronck one day, demanding to see Andy and wanting to know why he wasn't on the ward where he belonged. Lisa, the secretary, told her that at the end of January he had rotated onto another service and was no longer at the hospital. The woman demanded to know where he was, and when Lisa, who was well aware of the situation, refused to tell her, the woman pulled a big knife out of her pocketbook. One of the residents who had been sitting in the outer office ran to get the security guard who was stationed on the pediatric floor. The guard ran into the room, surprising the woman. In the confusion, he was able to overpower her and force her to release the knife. No one was hurt, thank God, and the woman was taken to the psych emergency room in handcuffs. She was ultimately admitted to Bronx State Psychiatric Hospital. Andy, who was working on the Infants' ward at

Mount Scopus, was relieved to hear this news, to say the least. He went home that night and had his first good night's sleep in weeks. And within a week or two, the whole incident was forgotten.

This melodrama is certainly not an everyday occurrence. But when something like this does happen, you can be sure it'll occur in January or February.

Andy

Sunday, February 23, 1986

All in all, the two weeks I spent in the Jonas Bronck OPD were pretty good, even though it was so frustrating. My prior ER experience at Jonas Bronck had been horrendous, and I had expected the same. But it was much quieter this time; the asthma room wasn't constantly packed, it wasn't constantly filled with screaming, wheezing children who were vomiting all over the floor, making the place smelly and sticky and making the whole emergency room so noisy because of the sound of the oxygen coming out of the wall tanks. Instead, it was much quieter, and on call nights we'd get out of there at twelve or one o'clock, instead of at four or five in the morning. And the chiefs, thankfully, were really nice to me for some reason. They gave me no Friday nights [**the night without a night float**] and the only even slightly hard thing I had was neurology clinic, which is bad only because there always are so many patients.

Even dealing with the ER staff was easier in January. I really felt like I was getting along well with the nurses for a change. When I had worked in the ER last there was this one nurse named Eve whom I didn't like at all. One day during my first month there I just said to her, "I've had it with you! I'm not going to ask you for any help anymore. All you ever do is give me a hard time! As far as I'm concerned, you're not even here! I'm not talking to you anymore!" And she said, "Fine." So we left on horrendous terms. She was in a really bad mood because she was going to be quitting at the end of November and at that point she hated being in the Jonas Bronck ER. And then one day I was seeing a patient in my clinic at Mount Scopus and I walked out of the examining room and there she was, there was Eve, whom everybody else loved and I hated. We were standing there, staring at each other eye to eye, and she kind of looked afraid. It was a strange thing; I had never seen Eve like that, she'd always been so nasty and aggressive. She had an almost scared look on her face. And I kind of just laughed and walked past her and said, "Oh, *you're* here!" And she said, "Yeah, I'm working here now, I'm one of the nurses here." And we both laughed, and she said, "Don't worry, I'm not going to be such a bitch because I don't know what I'm doing yet."

Things still weren't exactly great. But then, the next time I saw her, we talked for a little while and then we went out to lunch and now we've become friends. She's really a good nurse, she's fast and efficient, and she's funny. Now I even like her.

I should mention one patient I had in neurology clinic. He was a seventeen-year-old autistic, severely mentally retarded, violent guy who had been sent over from Bronx

Developmental Center [a residential facility for moder-
ately and severely developmentally disabled children and
adults] for evaluation because he was becoming increasingly
depressed and had been losing weight. He was on all kinds of
phenothiazines [a class of tranquilizers], but nothing was
helping. I brought him into the examining room with the
health care worker who had come with him. While I was
looking through his old chart, he suddenly started to become
extremely agitated. He got up and began stomping his feet
on the ground and then he started slamming his head into
the green tile wall over and over again. I looked over and
realized there was blood on the wall, and there was blood
coming out of his mouth. He had actually knocked a couple
of teeth loose!

Then he started going completely wild; he attacked the
health care worker and turned around and slugged me in the
ribs a couple of times. Then he went and smashed his face
against the wall again. The health care worker, a large,
matronly black woman, grabbed him and tried to hold him
still. All this time, he was screaming and making unintelligible
grunting noises. Pretty soon he began flailing around, and
the health care worker, who was getting kind of panicked,
looked over and said, "He doesn't like to be in tight spaces.
We need to get him out of here!"

I thought I had been very calm up until that moment. I
opened the door, walked out, and everybody was looking
toward the door because there had been a lot of ruckus in my
room. I said, "I have a problem here." We got the kid out of
the room and into the hallway, where apparently he didn't
feel so enclosed, and we called security and I sat down with
the neurology attending to whom this guy had been referred
and said, "Why did you put me in that room with him? You

knew exactly what was going to happen." And she made some comment like, "Well, you have to learn to take care of these kinds of patients," and I thought, No, I don't. As a resident, I don't have to take care of severely disturbed, autistic, retarded, violent patients who've been referred to a specific doctor for evaluation. In fact, the neurologist didn't want anything to do with him either. She finally said, "You know, we just can't evaluate him." So we sent him back with a letter saying sorry, there's nothing we can do. It was too bad, but we had nothing to offer.

I've thought about that patient a few times since this happened. I'm wondering how I would have reacted to the whole thing if this had happened back in July instead of in January. I think in July I would have tried a little harder, maybe looked farther through the chart or pushed the attending a little more. I don't know, the kid was crazy, he was dangerous, but when I was in OPD last July, I did some things for patients I don't think I'd do now: I stayed late to finish the workups on patients, things like that. I don't know if that's bad; it's just that it's a real change in me.

Anyway, those last weeks in January were very nice and I began regretting my decision to leave the Bronx. And on my last day in the ER, I said to the nurses on the afternoon shift, "This is the last time I'll ever be in Jonas Bronck." (My schedule at that time had been set so that I spend the rest of the year at Mount Scopus.) And they said, "What are you talking about? You'll be back next year." I said, "No, I won't." And they all said how much they'll miss me and stuff. It was very nice. Very nice and very sad.

I just finished my month in the ICU at Mount Scopus. It was a terrific month. We had a great team: Alex George was the attending, Diane Rogers was the senior resident, Terry

Tanner was the junior resident, and we had a couple of good cross-coverers at night. The ICU was tremendously exhausting but somehow I didn't feel as overwhelmed as I had in the NICU, where I never got any sleep. There were a few nights this past month when I didn't get to bed either, but I got at least some sleep most nights I was on call. And I slept for seven whole hours my last night.

I seemed to have the same luck in the ICU that I've had all through the rest of the year. I seemed to get the sickest patients with the most dismal prognoses, and Terry got a lot more of the acutely ill, rapidly recovering patients with relatively good prognoses. It became kind of a standing joke that if I were there to admit a patient, the patient would either wind up dead or with some kind of severe permanent deficit. I guess I've got a lot of bad luck.

There were three patients who were the saddest patients I'll remember for a long, long time. The worst was Ronnie Morgan, this wonderful, beautiful, redheaded boy. When I met Ronnie Morgan for the first time, he was intubated, with a shaved head, a swollen face, and a dozen lines running in and out of his comatose body. Ronnie Morgan was a little two-and-a-half-year-old who had been doing really well until three months before I met him, when he became ill with some minor symptoms and was found on a routine blood test to have an outrageously high white blood cell count. He was admitted, a bone marrow biopsy was done, and a diagnosis of ALL [**acute lymphocytic leukemia**] was made. Soon thereafter, he had a bout of ARDS [**adult respiratory distress syndrome, a condition in which the lungs fill with fluid and respiration becomes extremely difficult**] and a systemic fungal infection. He was admitted to the ICU at death's door, recovered, went through some chemotherapy, and finally was thought to be going into remission. Although his disease and his chemotherapy turned him into a cranky and irritable little kid, his mother always remembered him as being a beautiful, wonderful, happy boy. And then a few days before I first saw him, he was leaving his hematologist's office after a routine visit and fell and hit his head. That's not so unusual; he was a toddler, and toddlers fall a dozen times a day; that's why they're called toddlers. But when he fell, he happened to have a very low platelet count because of the chemotherapy, and he got an occipital hematoma [**a large, blood-filled bruise in the back of his head**]. So he was admitted to the hospital for a transfusion of FFP [**fresh frozen plasma, a blood product containing the elements of the blood essential for blood to clot**] and platelets because it was feared he could bleed out into the hematoma.

Over the course of the next thirty-six hours, he became progressively more lethargic, his mental status deteriorated, and with that concern he was brought down for an EEG [**electro-encephalogram, a test to examine brain waves**]. While he was in EEG, he began to seize. He was then rushed to a CT scan, where a massive intracranial bleed was found [**intracranial bleed: a hemorrhage in the skull**]. At that point, he was immediately brought to the operating room for emergency neurosurgery. There, a huge intraparenchymal [**within the body of the brain**] and subarachnoid [**below the inside layer of the meninges, the tissue that surrounds and protects the brain**] hemorrhage was evacuated, along with a good part of Ronnie's brain, something that happens when you do that kind of surgery. He was then brought up to the ICU on a ventilator and became my patient.

I knew Ronnie was a goner from the moment I saw him. He had a horrible problem, a subarachnoid and intra-parenchymal bleed, and that diagnosis on its own was horrendous. And then you add to it his age and his fragility with his leukemia and the low platelets and all the rest and he really had no chance at all. And everyone in the ICU identified so much with him and his parents, who were young, white, middle-class, articulate people.

He was with us for about ten days. After maybe the fourth or fifth day, he had a sudden, uncontrollable rise in the pressure inside his skull. We had been able to keep the intracranial pressure down prior to that time with various maneuvers and drugs, but that day it just became uncontrollable. And with that it was felt that he was essentially brain dead, and yet his body wouldn't die. It was all so horrendous, continuing to take care of this boy who had no prognosis at all. His father understood the situation; he knew how bad things were, and he was trying to mourn his son's death before it actually occurred. But his mother was too defensive and wouldn't accept it, and in a sense was preventing the father from doing his mourning. I never really got to know these people very well; when I first picked up Ronnie as a patient, I saw all the people who were gathered around the parents and I felt that my availability as a support person would not be needed. I didn't see the need of intruding myself into these people's lives when they had already made acquaintance with Alex George, a hematologist, and some other members of the staff. And while it was true that there was nothing extra I could have offered them, I think I missed out on something and I wish I'd had the opportunity to learn how to help these people grieve.

The day his ICP [intracranial pressure] skyrocketed, we

were standing by his bedside, and Alex said that Ronnie had died, that the bleed inside his head and the pressure had completely destroyed his brain. In the bed right next to him, separated only by a flimsy curtain, was a fourteen-year-old girl who had been diagnosed with a horrible brain tumor and who was going to die of that tumor within the next year or so. She had just undergone some surgery and she was a little off the wall and her mother, who had been sitting here, suddenly said, "I can't take this, this is too much for me!" and left the room. It was just a little too close to home for her.

Finally Ronnie did die. He died a couple of hours before I came in one morning. I had been post-call the day before and he'd been doing very badly; I left knowing it was only a matter of time. I didn't go to his funeral, but I wish I had. I think I've been mourning his death ever since he died; not often or always, but whenever I think about him, I get very sad. But it's strange: I never did know him as a person; I only knew him through the eyes of the people who loved him. Still, I know I'll remember him and be sad for him through maybe the rest of my professional career.

I had another patient, Kara Smith, a little four-month-old who broke my heart. Her first three months of life were normal, and then she came down with pneumococcal meningitis [**meningitis caused by the bacteria** *Streptococcus pneumoniae;* **this type of meningitis causes particularly devastating effects**]. When I picked her up at the beginning of the month, she was just this little seizing baby who was in renal failure, on peritoneal dialysis [**a procedure performed on patients in renal failure; dialysis rids the body of the waste products of metabolism that normally are removed by the kidneys**], getting multiple antibiotics, and who had a very abnormal neurologic exam. To make a long story short, Kara was the patient who should have died but didn't. Her mother agreed to a DNR order. I actually first brought up the idea of DNR with Ms. Smith. I told her it was something she should consider, and she decided that it would be best if Kara just died. It sounds cold-blooded saying it like that, but it really would have been the best thing that could have happened. We decided we'd just do supportive care and nothing heroic, but even that division became increasingly unclear.

One afternoon she had an acute respiratory attack; it was probably just mechanical, just her **[endotracheal]** tube slipping down her right mainstem **[bronchus, the breathing tube going to the right lung],** and she began deteriorating. She became cyanotic **[blue],** and her heart rate dropped. We readjusted her tube and she recovered a little but she still looked shitty. I felt really uncomfortable doing nothing, and yet, to bag her back was kind of a resuscitation. It was very unclear. And there was a senior, Eric Keyes, sitting there, and he said, "Forget it, just leave it, don't do anything." It was so easy for him to say that, it was nothing for him, he was just looking at her and thinking, This kid's just a GORK, forget her. But I felt bad doing nothing. She was my patient. I'd made a pact, in a sense, to support her. So when she continued to deteriorate, I in fact did bag her a little. And then, after a while, she was fine again. I'm still haunted by that issue. It's one of those very gray areas of medicine.

I talked to Karen so many times about this baby. One night after a really long discussion, she said, "Listen, you have no medical basis for what you're doing with this kid. Why don't you just put down the side rails on the bed and let the baby fall to the ground and die? Is that any different from letting her electrolytes get out of whack because you're not doing blood tests on her? Why not just put down the rails?" In fact, putting down the rails wouldn't make a difference; the baby never moved anyway. But she was right; there was no medical basis for so many of the decisions that were made. Everything seemed so arbitrary and based on emotions rather than facts. I think we were very much guided by the fact that the mother just didn't want that baby anymore, she didn't want a baby who was so severely damaged. We consulted the mother at many points along the road, we involved this poor woman who had no support system and who had other kids at home to worry about, and these decisions were difficult for her to make. One day I asked her to come in because we needed to talk about something; she showed up late at night, hours after the time I had asked her to come. She came with this young guy, about seventeen, who looked scared and a little intoxicated, and it turned out to be the baby's father. I had never seen him before; he had been nowhere near the hospital for at least two and a half weeks. We sat in the family room for a long time and talked; we talked about DNR and the new thing we were going to withdraw and she said, "Go ahead." It was as if she were saying, "Please don't let this baby live!"

It didn't really hit me how dismal Kara's prognosis was until I saw her CT scan. Her brain looked like a minefield; there was more space and fluid than there was brain tissue. This little baby had so little substrate to build her life on. But by then it was too late, all the miracles had been done on her; she just wouldn't die. I eventually got her off the ventilator and off the dialysis. I got her electrolytes corrected, I got her to feed, we stopped the antibiotics, and she became just a baby in a basinette. And one day we needed a bed in the unit and she was the most stable patient, so she was shuffled off to the Infants' ward.

The third patient was Emilio Diaz, a really adorable three-year-old with AIDS who had done so well for so much of his life. Emilio spent his first year in the hospital because there was no place for him to go. He was finally adopted by one of the nurses who had taken care of him when he was a baby. This woman really loved him. She married Emilio's father, an IV drug user who had AIDS so that she could legally adopt him and take care of him. She had done a lot to try to give Emilio as full a life as possible. He'd gone to Puerto Rico to visit relatives, he'd gone for trips all over the place, he'd done a lot more than your average Bronx three-year-old. And then he became sick and was admitted to the ICU and rapidly deteriorated. He had terrible pneumonia; he became ventilator-dependent and reached ventilator settings that nobody had ever seen before in the ICU. And he just kept getting worse until finally I had to have that horrible discussion with the parents about DNR on him, too. In Emilio's case, though, it was his parents who first asked about DNR. I suppose it's not surprising, since his mother is a nurse and very medically sophisticated. But one day she came to me and said, "He's suffered enough; we want to make him DNR."

Emilio died one evening just after I left for home. He was there the longest of all three of these patients; he was my patient for a good three and a half weeks. It was so sad taking care of him. There finally came a point, around the time that he officially was made DNR, when I felt very, very depressed. Every time I had to go over to his bedside, every time I had to write a note in his chart or I had to look at his bedside clipboard or call for his lab results, I became severely depressed. It was really sad. Really sad.

Well, anyway, I learned a lot of medicine from those three patients and from the other patients I took care of in the ICU. I learned how to put in A lines [**arterial lines; like IVs, except going into arteries instead of veins**], I got good at intubating, and I learned about Dopamine and Dobutamine drips [**a class of drugs known as pressors, which raise blood pressure; they're used in critically ill patients who cannot maintain their own blood pressure**]. And I did a lot of thinking about ethics and the fact that, basically, we can keep just about anybody alive for an indefinite period of time but that keeping people alive may not be such a good thing. That's a hell of a lot to pick up in one month! And I think one of the reasons I was able to do all that was because of the people I worked with.

Diane was a great resident. She's very, very bright; extremely capable; and very talented. She has a very wry sense of humor. We were on call together five of my nine nights, and we had a lot of fun. I thought she was kind of attractive, too, but of course that was something I couldn't really tell her. Her body is very similar to Karen's, and I found that very erotic. I told her that the morning after the last night of the month we were on call together. I was kind of delirious and I don't know how it came up but I said something like, "You know, you're very erotic," and she said, "What's wrong with you? You must be completely out of it!" And I said, "Yeah, I guess I am," but in fact I had been thinking about it for a while. But I couldn't say it until then because we'd been on call and slept in that little on-call room so many nights together. Not that it would have mattered, for God's sake. Anyway, it would have been an inappropriate thing to have done.

Alex is a great attending, and he really was very supportive. He'd come up, give me these big bear hugs during the day, and ask me how I was doing. He said he was worried about my psyche all the time because I was taking care of all these very sick patients. He was a good teacher, too. And he has a really big heart, he really cared about so many of the patients. He was somehow able to be a very devoted and involved ICU director, there all the time, always available and really involved, and yet he could keep some distance and let the residents try to run the place.

On the last day of the month, Alex took us all to lunch, which no attending has done for me before. He was a role model for me. I felt maybe I could actually do this kind of thing for a living. Even though it was depressing and even though taking care of critically ill children is so far removed from what I originally thought I'd wind up doing, I felt having someone who really cared made a big difference. When you think about it, all you really need in the ICU is a good technician who knows how to run all the machines and monitors. You don't *need* an Alex George to run a unit. But I think his heart makes a big difference; Alex is what makes that place seem human and not just a mechanical torture chamber. I'm sure his being there has helped a lot of families. I know it really helped the house staff; his fatherliness, his caring attitude, it really made a difference.

And finally, there were the nurses. I really got to like those nurses. They're superb; they're fun to work with, they've got a great sense of humor. I learned to depend on them totally and to trust them. They do a lot of things that need to be done without you ever knowing about them. They ask you to write the orders for them after they've finished. They know so much, they're each like a doctor. Many times we'll be sitting there, scratching our heads, trying to figure things out, and they're flipping the dials, running the lines, drawing up meds, making decisions they are confident with. And they just say, "This is what you want, isn't it?" as they push it in the line. So they are real lifesavers, and I'm going to miss them.

Amy

Sunday, February 2, 1986

All year, anytime anybody's asked me to do something, I've done it without an argument. It seems it works only one way, though, because whenever I've asked someone to do me a favor, nobody's willing to help out. It's disgusting.

This is what happened: I started my two weeks in OPD last Monday. I had only two weeks to go before my vacation, so I figured it would be a cinch. But when I got home from work last Monday afternoon, Marie told me she thought Sarah was coming down with something. She hadn't eaten well during the day, which isn't like her, she'd been kind of sleepy, and she had a runny nose. Sure enough, I put her to bed at about nine and she woke up at a little after eleven, screaming at the top of her lungs. Larry and I ran into her room and found that she was burning up. I took her temperature: It was a 103.4°. It took about twenty minutes to get her to stop crying, and when she finally calmed down, I checked her over. I couldn't find anything specifically wrong, her ears looked fine and everything, but she still had rhinitis [**runny nose**] and she was coughing a lot, and I noticed the whites of her eyes were red. I figured it was just the virus that was going around. I gave her some Tylenol and rocked her back to sleep, and she finally dropped off at about midnight.

She woke up again at 2:00 A.M. She was screaming, and her temperature was back up. I gave her more Tylenol and tried to calm her down, but this time she just wouldn't stop crying. I was sure she had meningitis and I told Larry to get dressed because we had to take her to the emergency room, but just as he finished getting his clothes on, she quieted down and fell back to sleep. I guess the Tylenol had kicked in. Anyway, she slept the rest of the night, but I didn't; I stayed awake in her room, watching her constantly. I was sure something terrible was going to happen.

On Tuesday morning, she woke up in a much better mood and her fever was gone. I figured whatever was wrong had reached its peak and now she was getting better. When Marie came, I told her about what had happened and made sure she knew how much Tylenol to give if the fever came back. I had clinic that morning and was going to be in the ER that afternoon and night, so I left Marie a schedule of where I'd be if she needed to contact me, and I left for work.

I should have called in sick, but I went to clinic anyway. Marie called at about ten-thirty to tell me that Sarah's fever was back and that she had this rash all over her. I told her I'd be home in a few minutes and I rushed through the rest of my clinic patients. I was done by about eleven and I ran home to find Sarah's fever back up to 103. She was miserable; she was coughing and sneezing and covered with snot, and she had a whopping conjunctivitis. And she had a raised red rash on her face and chest. I wasn't sure what it was, so I called Alan Cozza. He told me to bring her right over.

Well, to make a long story a little shorter, Alan took one look at her and said, "My God, she's got the measles!" I had never seen anyone with measles before; kids just don't get it, because we immunize them. Alan brought some of the other interns who were on the floor in to see Sarah just so they'd know what measles looked like. I have no idea where she got it; she's a baby, she doesn't go outside, she doesn't hang around with other kids except sometimes when Marie takes her down to the lobby, but that's rare. But anyway, she had it. Alan told me to take her home and give her Tylenol and fluids and just make her as comfortable as possible and that it would pass in a few days.

By that point it was nearly one o'clock. I was supposed to be in the ER starting at one, and since I was on call that night, I'd be staying in the ER until maybe three or four in the morning. So I decided to stop in and talk to the chief residents; I figured, hearing that Sarah was so sick, they'd naturally say, "Well, why don't you just stay home with your daughter tonight?" Yeah, right!

What they said was that two people who were supposed to be on call that night had already called in sick and they had to pull one person from the emergency room to cover and although they sympathized with me, they just couldn't let me off. If I didn't show up, there'd only be two house officers to staff the entire emergency room, and they just couldn't allow that to happen. They told me I should try to switch with someone who was scheduled to be on the next day, if I could find someone who would be willing to switch, but there was just no way they could give me the night off.

How nice of them! After all the abuse I've taken through these seven months! After everything I've done for them! Whenever they've asked me to do anything, I've always done it without a whimper! I filled in for other people, I covered wards I'd never been on before because somebody was out, and I never complained. I've repeatedly put my job ahead of my family, and this is the thanks I get! The one time my daughter is sick, the one night I need to take off, of course no one would do a thing for me. I asked everybody if they'd switch, if they'd cover for me this one night, and they all had some excuse. I should have just gone home. I should have taken Marie and Sarah home and stayed there and when they called me to find out where I was and why I wasn't in the emergency room, I would have said . . . I don't know exactly what I could have said. But, of course, I didn't do that. What I did was, I brought Sarah and Marie home and went back to the hospital.

The rest of that day was ridiculous. We were short-staffed in the emergency room; the place was like a zoo. Everybody in the Bronx was sick with the flu and had fever and vomiting and coughing. There was a six- to seven-hour wait to be seen through most of the afternoon and night. I rushed around that emergency room until four in the morning, and during all that time I didn't see a single patient who was as sick as Sarah. I really resented being there, and I must have told that to the nurses and the rest of the staff at least a thousand times.

Marie kept calling me all through the afternoon. She didn't exactly feel comfortable taking care of a baby with a fever that ranged between 103 and 104, and I can't say I blame her. I'm sure she was afraid Sarah was going to have a seizure or something. That idea crossed my mind a few times.

So she called every half hour or so, saying, "Her fever's still up. What should I do?" or "Her eyes are getting very glassy. Are you sure you can't come home now?" Even she had a hard time understanding why I couldn't just come home to take care of my baby. And there wasn't anything I could say to her to make her understand because I wasn't sure I understood it myself.

I finally got out of the emergency room at about four. Larry was wide awake; he hadn't gotten any sleep, having been up with Sarah all night. Her fever was still up, and she was very irritable. She'd sleep for maybe a half hour and then wake up howling. It's so strange seeing her like this; she usually has such a good personality. And she was absolutely covered with the measles.

I was all set to call in sick on Wednesday, but Larry had already made arrangements to take the day off, and he told me I should go in. I went, and in the morning Alan called to ask me how Sarah was doing. I gave him a piece of my mind! I told him about how the chiefs had made me work the night before, and he seemed amazed by it. He said he'd go have a talk with them, but a lot of good that's done me! I'm really so angry. I'll tell you, this episode has really taught me a lesson. Let them ask me to do anything, let anybody ask me to cover or to switch; my answer is going to be *"NO!"* I don't care what it is or who it is, I'm not doing anything for anybody ever again! I've had it with these people! I've got to look out for my own interests, because no one else is!

Sarah's better now. The rash is starting to fade. Her temperature came down to normal on Friday, and she's not irritable anymore. By tomorrow she should be back to her old self. But I'm not going to forget this. You can bet they're going to regret making me work Tuesday night!

Saturday, February 8, 1986

I've just finished packing. My vacation starts after I finish my call in the ER tomorrow night, and we have a ten-twenty flight to Fort Lauderdale on Monday morning. I really need this vacation. I'm physically and emotionally exhausted. We're spending two weeks in a condominium near Fort Lauderdale; two weeks of lying in the sun and sleeping late. I can't wait!

Things have pretty much returned to normal. Sarah is back to her old self. The measles have disappeared, and the only things left from the whole episode are my anger and resentment.

Mark

Monday, January 27, 1986

Today was my first day in the nursery. What fun I had! I love staying until ten-thirty, running around like a chicken with my head cut off, having no idea in the world what the hell I'm doing. It was a million laughs! I can't wait to go back there tomorrow!

Well, let's see: How can I describe what the day was like? I came in at about eight o'clock, and Ed Norris, the director of the neonatal ICU at Jonas Bronck, tried to give us an orientation lecture. He certainly made things very clear; it was like listening to a lecture in Swahili! I couldn't understand a good 50 percent of the things he was talking about. He kept referring to the inhabitants of the unit as "your patients." You know, "your patients this" and "your patients that." And then we walked around and he showed us these so-called patients. My God, those things weren't patients! They'd have to quadruple their weight to be classified as patients. Right now, they're mostly tiny portions of buzzard food with lots of ridiculous wires and tubes coming out of them. This is going to be a long month!

They gave me eight of these things to take care of. For most of them, the kid's chart weighs more than the kid does, which is a very bad prognostic sign. And, of course, I didn't have a clue about what the hell was going on with any of them, so I spent the whole day sitting in the nurses' station trying to read these ridiculous charts, which were filled with words I had never seen or heard before and numbers I couldn't even attempt to figure out. I was trying like hell to make sense of all of this before one of these things wound up dying. At least I didn't have to go to clinic this afternoon; it was canceled. Thank God, because there's no way I would have made it anyhow.

I really can't believe this! It's ten-thirty, I'm just getting home from work, and this is my good night! It's just amazing! I have eight patients, they're all stable, and none of them is really that complicated, but even uncomplicated preemies have this long, annoying history, most of which I don't give a damn about. I mean, truly, I just don't care! There was a point there, about five o'clock tonight, where I swear I was this close to just taking all the charts, throwing them out the window, and saying, "Forget it! I'm sorry I ever applied to medical school! I never really wanted to be a doctor anyway!" I just couldn't take it anymore: just all these little runts who shouldn't be alive in the first place! Damn! Really annoying! But hey, I stuck it out, because I have such great self-control, and here I am, celebrating by eating my favorite food, Sno-Caps. This is the first thing I've placed in my mouth since breakfast. That was over fifteen hours ago! Working in this damned ICU is like being on a self-imposed fast! I feel like Gandhi, for God's sake!

I better try to eat neatly. This looks like it's going to be one of those months where I'm not going to get to wash the dishes or do the laundry! Maybe I'd better just start using paper plates right now. It's too bad they haven't invented paper clothes. That'd be perfect: disposable clothes for the house officer. Maybe if they could be made edible, that would solve both problems at once. I'm not making any sense anymore. I've got to get some sleep!

This place really sucks. What am I going to do?

Wednesday, January 29, 1986

Well, I'm home again. It seems like only yesterday I was last here, but actually it was the day before yesterday. I'm exhausted. I was on last night and I didn't get any sleep. No one ever gets any sleep in the nursery, so saying you didn't get any sleep on a night you were on call is redundant. I spent the whole night running around from bed to bed, doing stuff I didn't understand on babies I didn't think were human, for reasons that are totally beyond me. What a rewarding experience.

Here are a few more of my thoughts about the neonatal intensive-torture chamber. What a fun place it is. Starting to work in the NITC **[neonatal intensive-torture chamber]**, or the NICU, as the neonatologists like to refer to it, is like being thrown into prison in a foreign country where you have no idea what the fuck's going on. I really don't know anything! I've never even taken care of well babies before. I can barely tell the difference between the respirators and the babies. Before Monday, I'd never seen a kid with jaundice. **[Neonatal jaundice is caused by immaturity of the liver,** the organ that removes bilirubin from the bloodstream; it's a common problem in infants and is treated with phototherapy, placing the infant under banks of fluorescent lights.] I don't know when to turn on the lights, when to turn off the lights. I don't even know what a normal bilirubin is for a baby!

And I don't know what you're supposed to feed these things. I don't know how much they're supposed to eat, how much they're supposed to pee, nothing! If it weren't for the nurses, who, thank God, seem to know what the hell's going on, I'd probably have managed to kill off every last kid by now!

And even if I did know all that simple, obvious stuff, there's all this other information I don't even have a clue about. There isn't one word that they use in there that even sounds like anything I've ever heard outside the unit. It's like they make up terms just to make our lives more miserable, if it's possible to be more miserable than I already am! Every one of those kids has biochemical rickets. What the hell is biochemical rickets? I have no idea! And besides that, who cares?

As you can tell, all these little annoyances aside, I'm really having a lot of fun. I'm really enjoying taking care of these bags of protoplasm. My favorite patient is this kid Moreno. He's a three-month-old with congenital hydrocephalus. **[Hydrocephalus is a condition in which an excessive amount of cerebrospinal fluid, the substance that normally bathes and protects the brain and spinal cord, accumulates in the skull. Usually it is due to obstruction of flow of the fluid from the brain, where it is produced, to the spinal cord, where it is absorbed. When hydrocephalus occurs at birth, it is usually caused by an abnor-**

mality in the formation of the brain.] This kid is all head!
He weighs twenty-five hundred grams, and about twenty-
three hundred of those are housed above his neck. And of
those twenty-three hundred grams, 99 percent of that is
fluid. His cerebral cortex looks like a ribbon around a water
balloon. And that's after he had a shunt put in that seems to
be working. [**A shunt is a piece of plastic tubing, one end
of which is placed in the brain, the other end of which is
placed in either the abdominal cavity, the chest, or the
heart, that drains the cerebrospinal fluid out of the brain
in patients with hydrocephalus.**] His head circumference
today was forty-nine centimeters. [**Normal head circumference for a newborn is thirty-five centimeters. At three
months, the head circumference should be about forty
centimeters.**] This kid's got a great prognosis!

So anyway, Moreno's mother called me today. I wasn't in
a very good mood, having been up all night and not having
understood anything anybody has said to me in nearly three
days, so I wasn't really in much of a mood to put up with her.
She calls every few days to ask what the kid's head circumference is. She's fixated on his head circumference. I told her it
was forty-nine centimeters this morning and she got all panicky, saying it was only forty-eight centimeters on Sunday and
now it was a centimeter larger, and wasn't I worried about it,
and what was I going to do about it? I calmly explained to
her that no, I wasn't worried about it because it had been
forty-nine centimeters when I got there on Monday and it
was still forty-nine centimeters now and he wasn't irritable or
vomiting and he didn't have any of the other signs of
increased intracranial pressure, and since the neurosurgeons
and the neurologists had been by to see him and neither of
them had been upset by his head circumference, I wasn't

going to do anything about it. I think I also told her that I was happy that his head circumference was forty-nine centimeters, I was pleased as Punch, and if she wanted to find someone who wasn't happy, I suggested she call the neurosurgeons or Ed Norris to see what they think. I think I said that, but I'm not sure because, like I say, I was kind of tired and I haven't been making much sense over the past few days. But I'm pretty sure of one thing: I don't think Mrs. Moreno is going to be calling me much during the rest of the month!

Well, there is one saving grace about working in this torture chamber. Some of the night nurses are extremely cute. One in particular: dark, brown hair, really beautiful. Damn! She almost makes it worth staying up all night. But not quite. Nothing could really make it worth staying up all night.

I must say, my progress notes have deteriorated significantly. I never really wrote very good notes in the first place; in fact, my progress notes have been voted among the worst ever seen at Mount Scopus Hospital. Recently, no one's been able to read any of them. But at least they used to be short. Now, because of all the problems these kids have, instead of my usual three or four lines of unreadable scribble, I now write whole pages of unreadable scribble.

What a stupid thing to do to us, throw us in the middle of this unit when we don't know what the hell we're doing. And Norris screams at us that they're our patients! Bullshit! He should be thrown in jail if he really thinks they're our patients! None of us knows what the hell we're doing with them. All right, show us around, give us a week or two to figure out what's going on, then you can think of them as if they were our patients. At this point we can have virtually nothing to do with their care, because we know virtually nothing about how to care for them.

Hey, but the jury's not in yet. I'll give it a little more time to see what it's like before I make up my mind. I'm on call Friday night with a senior resident who sort of drives me crazy. She's reasonably intelligent and she seems to know what's going on, but she really lacks self-confidence. It wouldn't be so annoying if she didn't keep turning to me for reassurance. Me, can you believe that? I mean, I have absolutely no idea what the hell is going on! The other night she did something that I think was probably wrong. She wanted to intubate this kid for having one bad blood gas. And she asked me if I thought it was the right thing to do. I said, "No, it doesn't sound right. I think you should just turn up the oxygen and repeat the gas before you do anything." That was just common sense; this kid was perfectly pink at the time. I hope I'm not that unsure of myself next year, when I'm in a position of authority. Hah! Boy, we're all going to be in trouble when that happens!

Thursday, January 30, 1986

It's nine o'clock and I just got home. Things are looking up: I cut an hour and a half off the time I finished on Monday night. Why, if this keeps up, I'll have so much free time this month, I may actually get to cook my own dinner one night! I think washing the dishes'll still be out of the question, though. That'd be just too much to shoot for.

Here's some news on the Moreno front: Today's head circumference was forty-nine and a half centimeters, up one-half centimeter from yesterday's closing. And you know what? I don't give a shit! I just don't care!

I don't like the neonatal ICU. I'm not positive, but I

just don't think I'm going to grow up to be a neonatologist!

I've got this nervous feeling in my stomach all the time. I stopped at a drugstore on the way home and bought this great big bottle of Maalox. Either I'm coming down with gastroenteritis or I'm beginning to burn a big hole in my gastric muscosa.

Monday, February 10, 1986

I'm post-call. There's nothing like being post-call when you're in the NICU. It's at least a hundred times worse than being post-call anywhere else in the world, including Infants', which until this month had won the prize hands down as the Most Horrible Post-call Experience in the Bronx. I should be asleep now, but I haven't recorded anything in over a week. I've wanted to; I just haven't had enough strength to push down the "record" button on this silly machine. I don't want to let this fabulous experience escape immortalization on cassette tape, so, at great expense (at least ten minutes of precious sleep), here goes.

Working in this nightmare has now settled into a nice, regular, predictable routine of devastation and misery. Take yesterday, for instance. It was Sunday and I was on call. I walked through the doors of 7 South with a smile on my face at seven-thirty and was completely and overwhelmingly depressed by eight o'clock. Iris Davis, who'd been on call Saturday, signed out to me. She was in a great mood. When she gets real tired, she starts to cry, so it took her about an hour and a half and at least three boxes of tissues to get through sign-out. All I got out of it was a scut list about a mile long and a terrible headache.

Most of what Iris signed out to me was checking bloods that had been drawn earlier in the morning. So I started calling the labs to get the results, a very rewarding experience. Each lab had a different and very novel explanation for why the results weren't available. The chem lab claimed they had never received any samples, even though Iris assured me that she hand-delivered them. The hematology lab said that all the specimens, every one of them, was QNS [**quantity not sufficient**]. That's a polite way of saying, "We poured the blood down the sink, so you're going to have to draw them all over again." And the blood gas lab said the machine was broken and they wouldn't be able to run any samples for at least another hour. What this all meant was that I was going to have to spend the next two hours redrawing all these bloods and then spend another half hour delivering the samples to all the different labs.

Okay, so I did all that, and then I got to spend the next six hours writing progress notes. I love writing progress notes! You have to write a note on every patient every day or else the administrator on duty swoops down at about midnight and puts an evil spell on you. And these notes aren't just "Patient still alive. Plan: Make sure he stays alive until tomorrow morning." These notes go on and on, listing problem after problem. Each one can take an hour.

It wouldn't even be so bad if all I had to do was the scut and the notes, but that's all broken up by the endless rounding. First the attending showed up at about eleven so that he could view the patients close up for about an hour. He also has to fill his daily quota of yelling at the house officers. Then in the evening, there were rounds with the senior resident, who managed to come up with a whole new list of scut for me to take care of during the night. So I got to draw some more blood, have more fights with the lab technicians who have perfected the art of denying having received blood samples you handed them not an hour before, and write more long notes in the charts documenting what the results of those fights with the lab technicians have been.

And even all that would have been okay if it hadn't have been for the fact that the DR [**delivery room**] kept calling us to come down to deliveries. That's the real ulcerogenic part of this job. The rest is just irritating, but the DR is downright frightening. At any moment, without so much as a minute's warning, you could be called down there and find yourself face to face with a brand-new four-hundred-gram wonder whose only goal in life is to make your next two or three weeks completely unbearable. Also, there are all these little emergencies that come up in the unit, like kids deteriorating or spiking fevers, stuff like that. It's all such fun!

Last night, I actually got myself in a position to go to sleep at about three o'clock. I was in the on-call room, in my winter coat, getting ready to lie down on the cot. The on-call room is an interesting place. They recently rebuilt the entire ICU and they made us this very nice place to sleep. The only problem is, they forgot to put any heat in there. The average temperature is about forty to forty-five degrees. Sleeping in there is like camping out in Alaska.

Anyway, I was on the cot, getting ready to lay down. I was lowering my head toward the pillow and just as my hair made contact with the pillowcase, my beeper went off, calling me to the DR stat. I went running down there to find it was an uncomplicated problem, a little meconium-stained amniotic fluid. The baby was out by the time we got there and he was fine, just fine. There was nothing we needed to do. So I went back to the unit, got back into the on-call room, put my coat back on, and actually got about an hour of sleep. At five o'clock, I got another stat page. We went running down to the DR, and what did we find? An obstetric resident with her arm, up to the elbow, thrust inside a woman's vagina. What a romantic sight!

It turned out, this woman had wandered in off the street with a prolapsed cord. **[The umbilical cord had come through the cervix and was lying in the vagina. The danger of this is that if the cervix should close up again, blood would stop flowing through the umbilical cord and the fetus would suffer from lack of oxygen, causing either death or severe brain damage.]** The resident was trying to push it back up into the uterus while two other people were preparing to do an emergency C-section. The whole thing took about an hour, and when the baby came out, he was just fine! By that point it was eight o'clock in the morning and time for the day crew to show up. I finished with the work on that patient, started drawing the morning bloods, and then started rounds with the rest of the team.

We went on work rounds and then we had attending rounds. I tried to write my notes during all this because I had clinic this afternoon and I didn't want to have to come back to the unit again after clinic was over. So I got to clinic at Mount Scopus at about two and I got home a little while ago, at about six. A typical thirty-four-hour day; at least I got one whole hour of sleep!

I've had some really terrific experiences in the unit over the past week. Really terrific! I've got these two kids who are essentially brain stem preparations. One weighed 525 grams at birth and from the very beginning had virtually no chance of surviving. So what do we do? We use everything we have to keep him alive. And all that comes out of it is a great deal of work for me and the other interns. The other kid was good-sized at birth, a thirty-three- or thirty-four-weeker, but the mother had an abruption [**abruptio placentae: a condition in which the placenta tears itself off the wall of the uterus, leading to a great deal of bleeding and a severe deficiency of blood in the fetus**] and the kid was severely asphyxiated, with Apgars of 0, 0, 3, and 3. Not what you'd call very good. Where I come from, we have a name for children like this: stillborn. So this kid is basically brain dead but we're keeping the body alive to have something to keep us busy. As if I already didn't have enough to keep me busy!

This rotation continues to have only one redeeming feature, that being the nurses. These nurses are fantastic. They're young, real attractive, and real good at their job. Last week one of the night nurses handed me a prescription form. It looked like this:

CITY OF NEW YORK
HEALTH AND HOSPITAL CORPORATION

_____HOSPITAL

Name _*Mark Greenberg*_____ Age _26_

Address_____ Date _2/4/86_

Rx

l date with me

Physician's Signature _555-0826_____

I've been carrying it around ever since. I had lunch with her last week. She seems nice. I don't want to jeopardize my relationship with Carole; Lord knows it's already suffered enough! So I don't think I'll actually take this any farther. But it sure was nice to get that note. It made me feel . . . it made me feel almost as if I were a human.

Tuesday, February 25, 1986

I haven't recorded anything in a couple of weeks. I look upon these tape recordings as kind of a funny running monologue, but I haven't felt very funny over the past two weeks. Working in this unit has been terrible, just terrible, much worse than I ever imagined. We've had a lot of deaths and, even worse, we've had a lot of survivors, babies who should never have been allowed to live. I don't want to think back on what's happened, I just want to look ahead. In just a couple of days I'll be done with the NICU, and then I've got a month in OPD at Mount Scopus, two weeks of vacation, and two more weeks of OPD after that.

Every month so far there have been a lot of bad memories, but there have also been some good ones, funny stories. I'll carry with me probably forever. This month there have only been bad memories and worse memories. Moreno and his steadily increasing head circumference; the wasted, dying preemies hanging on much longer than they should be allowed to because of all the machines we have to use on them; the bigger kids with PFC [**persistence of the fetal circulation**]; the brain-dead baby with the abruption; these were terrible, terrible things. I've been finding that I just can't defend myself against them. It's been just brutal.

I don't want to make this sound too sappy, but I knew I was in trouble when I cried in the hospital last week. Iris, the other intern, has been crying just about every day, but last Thursday I had been up the whole night before with this PFC'er who had done really poorly, and when he finally died, I just couldn't take it anymore. I went into the bathroom, locked the door, and just cried my eyes out. I'm really starting to fall apart. That was the first time I'd cried all year. I know most of the other interns have cried, but I kind of prided myself on the fact that I could control myself. Not this month.

Maybe sometime in the future I'll be able to come back to this and fill in some of the blank spaces I've left, but I can't do that right now. I need a nice vacation. I think I'll take my vacation in the West Bronx emergency room over the next four weeks.

I'm going to sleep. Maybe when I wake up, things'll start being funny again.

Bob

Mark Greenberg and Andy Baron worked in ICUs during February, and both had experiences caring for patients who were being kept alive thanks to technological advances that had been developed over the past few years. It's always been true that technology has run way ahead of ethics in medicine. With every advance that's been made, be it the development of antibiotics, the iron lung, present-day respirators, chemotherapy and radiotherapy for cancers, or the ability to transplant organs, physicians have been able to take discoveries made in the laboratory and apply them to humans. The immediate result of these advances has been that patients who the week before would surely have died have been given the opportunity to survive, at least for some period of time. But we've often learned that survival may not always be the best outcome for the patient or for society. The question of whether these fruits of medical technology should be utilized has to be addressed. In many cases, answering this question can be more difficult than developing the technology in the first place.

In no place is this truer than the neonatal intensive-care unit. Although there has always been interest in the very premature baby, until the 1960s these infants were considered little more than curiosities. Rather than being cared for in specially designed intensive-care units where all their life functions were meticulously monitored, these babies used to be warehoused in circuses and freak shows, and exhibited to the public for a price. If they lived for very long, it created more interest. If they died, usually because of respiratory failure, it meant only that they needed to be replaced.

Unlike most other medical specialties, which gradually evolved into existence, neonatology had a sharply demarcated beginning, largely the result of a specific event. In August 1963, Patrick Bouvier Kennedy, the premature son of President John and Jacqueline Kennedy, died of respiratory distress syndrome at Boston Children's Hospital. For a few days, an intense media spotlight was shone on the special problem of babies who were born too soon. Although, sadly, this event ended with the death of the infant, it resulted in millions of dollars of national grant funding being devoted to research into the special problems of the premature. And therefore the death of Patrick Bouvier Kennedy led to neonatology as we know it today.

The major advances in the field occurred early. By the mid-1970s, using respirators, intravenous medications and fluids, specially developed dietary formulas, and very aggressive care, it became technically possible to keep alive infants who were born as much as fourteen weeks prematurely and who weighed as little as twenty-eight ounces. Some of these infants did well; they've gone on to lead relatively normal lives. Most of the other survivers, though, have been left with significant physical and developmental problems: Some developed cerebral palsy and required orthopedic intervention and braces to help them walk but were otherwise spared; others were found to have suffered extensive brain damage and, in addition to cerebral palsy, were left with mental retardation and seizures; still others were so extensively damaged by the consequences of their prematurity that they wound up leading a vegetative existence, many residing in institutional settings. And so the question was raised, "Although technically possible, is any of this justified?" Neonatologists and medical ethicists have been struggling to answer this question ever since.

Neonates with problems can be divided into three groups. A first group includes those who have an excellent prognosis right from the beginning. This group includes the "garden variety" preemie who weighs two pounds or more and who is born without any problem other than prematurity. Most everyone in the field of neonatology would agree that everything possible should be done to support these infants.

A second group is made up of those infants who are born with such severe defects that survival is not possible no matter what is done for them. Included in this group are babies with anencephaly, a condition in which the skull and brain fail to develop; all of these infants are either stillborn or die within the first days of life. Also included in this group are babies born before twenty-four weeks of pregnancy. Most but not all neonatologists feel that these infants should be made as comfortable as possible and be allowed to die without intervention.

The third group of infants with problems is the most difficult ethically. It is made up of those children who fit between these two extremes: babies born weighing less than two pounds but above the twenty-fourth-week-of-pregnancy cutoff; and infants with major birth defects that are not necessarily lethal. The medical community is divided about what to do with these babies. Many neonatologists would do everything possible to offer these infants the opportunity to survive, knowing that possibly for every surviver who turns out to be normal, there'll be an infant or two who will wind up significantly damaged. Others would provide limited care, reasoning that the "strong" will survive and the "weak" will die off (the problem with this reasoning is that some in the former group who would have led a normal existence had aggressive care been provided will wind up damaged as a result of this method). Finally, some would argue that nothing should be done for this middle group and that nature should be allowed to take its course; physicians who think this way are clearly in the minority.

In neonatology, there's a tendency to lose sight of the end point. Sometimes a neonatologist who understands that he or she has the tools to keep any newborn alive for as long as he or she wants, may decide to flex his or her technological muscles and play God, keeping alive children who should be allowed to die. Neonatologists might argue that these exercises are good in the long run: By learning about keeping these children alive even for a brief period today, it might someday be possible for some to survive. And there might be some truth to this; after all, the argument that nothing should be done could have been made twenty years ago concerning babies who weighed twice what the babies who survive today weigh. But the question is, What price is being paid for this?

During his month in the NICU, Mark was kept awake night after night caring for babies some of whom he considered brain dead, one of whom weighed only a little over seventeen ounces. Discouraged because of all the inevitable deaths, he asked, "What possible good am I doing here?" It didn't seem as if many of the babies were benefiting from the intensive care. The parents, who were seeing everything done for their infants, were being given false hope; they reasoned, "If they're doing so much, they must believe that my baby has a chance to survive." This winds up making coping much more difficult for the parents when the baby ultimately does die.

Taking care of patients who have no chance of surviving is extremely frustrating and anxiety-provoking. You're asked to do things that don't make sense to you; you're called upon to counsel parents without having the picture clear in your own mind. But this state of mind is not limited to working in the NICU. These problems also occur in other intensive-care units.

Physicians working in ICUs that care for older patients must deal with many of the same issues as the neonatologist, but the situations are often radically different. Patients in the pediatric or adult intensive-care units are not neonates; they come into the unit with a life history. They have relatives and friends who know them and love them, not just for what they might be in the future but also for what they've been in the past. They have personalities and desires, and often specific requests about what should and should not be done. The intensivist must often decide whether to honor these requests, or the requests made by the patient's loved ones, or to do whatever he or she thinks is in the patient's best interests. And that can be very difficult.

During his month in the pediatric intensive-care unit, Andy became involved with three patients for whom "do not resuscitate" orders were ultimately written. Actual orders stating that a specific patient should not be resuscitated in the event of a cardiac or respiratory arrest are new at Mount Scopus Hospital. Prior to the time that the present interns began their year, plans for patients who had no chance of survival were formulated through conversations among the physicians, the family, and, if possible, the patient. If it was agreed by all parties that resuscitation should not be attempted, the word would be passed to all members of the care team. The concept of a patient being a "no code" developed; then the concept of a limited or "slow code" (the situation in which cardiac arrest leads to limited efforts at resuscitation) evolved. Verbal "no codes" were troublesome; to many members of the care team, it seemed like a sham. Notes and orders were being written in the patient's chart that did not truly reflect the thoughts of the care providers or the wishes of the patient and his or her family. But DNR orders could not be written; they raised legal and ethical questions that had not yet been answered.

The use of written, formalized DNR orders arose through the efforts of a committee composed of the hospital's lawyers, ethicists, and physicians. Now the true plan for a specific patient can be spelled out in the chart without fear of legal or ethical retribution. The actual order must be written by the patient's attending physician and must be reordered every week. Once a DNR order has been written, it leads to conflicts of another sort: Now that we've stated that the patient is expected to die, what should and what should not be done for that patient?

Here's an example that'll help explain this conflict: A

patient who is DNR develops a fever. Normally, hospitalized patients who develop fever are managed very aggressively; a "sepsis workup" consisting of blood and urine and sometimes spinal fluid cultures is done, and antibiotics are immediately begun. Failing to treat a patient with fever may lead to overwhelming infection and ultimately to death. But what should be done if the patient is DNR? Should antibiotics be started on such a patient, or should infection and its consequences be "encouraged"? If antibiotics are going to be withheld, should cultures be obtained? These questions must be considered in every case. Often, under the reasoning that to treat an infection would be to prolong life artificially, antibiotics will not be given and cultures will not be obtained.

But using this reasoning, one could argue that feeding the patient would also lead to artificial prolongation of life. Therefore, should DNR patients receive the nutrition they require for life to continue, or should they be allowed to starve to death? Most physicians would agree that the withholding of nutrition should not occur. Implicit in DNR is that the patient should not be allowed to suffer. Starvation is a painful and drawn-out way to die. Therefore, most intensivists would make sure all patients were receiving an appropriate number of calories to sustain life.

These are only some of the issues Andy and Mark agonized over during their month in the ICU. And they are not alone. The conflicts that arise for young physicians at the edge between life and death are universal. And they lead to a great deal of mental and emotional stress and anxiety.

Andy

Tuesday, March 11, 1986

For the past two weeks I've been in the OPD at Mount Scopus and West Bronx. It really hasn't been too bad. I've come to realize that I've had to start acting more like a resident; I have to depend more on my own impressions and make my own decisions. The past few weeks have been the first time I haven't felt that the residents and the attendings were giving me good answers or helping me solve problems very well. So it's been a kind of stressful learning experience, but I think I've been doing okay at it so far. I guess this is how you learn to become a resident.

The other night in the Mount Scopus ER was memorable. It was my last official night in the Mount Scopus ER. I was supposed to have another whole month of OPD on the west campus, but I switched to be at Jonas Bronck. Working in the Jonas Bronck ER is a better learning experience. So it was my last on call; I can't say I'm not happy to get it over with.

It was also one of the worst nights I can remember. There was a tremendous volume of patients; they kept just coming in. It was nuts! At one point we were fifteen charts behind, which is a lot for that place, but we couldn't make any headway because we had about a half dozen acutely ill children. And the place has only four rooms; we were spilling over into the adult ER. Let's see: We had two head traumas in various states of coma; we had a diabetic with sickle-cell disease who was in the middle of a painful crisis *and* in DKA [**diabetic ketoacidosis, the buildup of acid in the blood of diabetics caused by high sugar in the blood and inability of the cells of the body to use the sugar for its normal processes**]; we had a little baby sickler with fever; we had a couple of vaginal bleeders and a drug overdose. All of these were occurring pretty much simultaneously. And it was just me and a senior resident who was not the greatest doctor you ever saw. We couldn't get help from anybody. The attending was over in the West Bronx emergency room. Every time we'd call with a problem, he'd say, "Well, it doesn't sound *too* bad. Call the senior in the house [**the resident in charge of the inpatient service at night**] if you're worried." He wasn't even concerned! What a shithead!

I never ran as hard as I did that night. Finally at one point the nurse, who was fabulous, said to us, "Please call for some help!" So we did. And then slowly but surely we got some of the docs who were on call on the floors down there, and we cleared the place out. But it was still crazy the rest of the night. Right before we were going to leave, this fourteen-year-old girl who's an asthmatic and has been intubated twelve times came in tight as a drum. [**She was not getting air into the lungs. Asthma is caused by narrowing of the air tubes. When these air tubes are slightly narrowed,** wheezing will be heard in the chest; when they become very narrow, as they were in this patient, no breath sounds are heard and the patient is considered "tight."] We had to intubate her in the ER. I didn't get out of there until 3:00 A.M., which is late for Mount Scopus.

Karen left a couple of weeks ago after nearly two months of that subinternship she was doing. It was sad taking her to the airport. Fucking LaGuardia Airport; I really hate that place! I've felt very blue since she left. I've been missing her a lot and it's been a real drag being apart like this. She's doing obstetrics/gynecology now; she has to be on call every third night, and our schedules are completely out of whack. We've been able to talk only a couple of times in two weeks. It's weird. But we'll be together soon. It won't be too long before this insanity is over.

Last Tuesday was my birthday. I was post-call and I felt terrible, and I didn't want to celebrate at all. I went to visit my friend Gary and his roommates out in Brooklyn. I had a good time. The next day I went out with my friend Ellen. We went into Manhattan and had a wonderful time. Anyway, the weekend was pretty good.

I've been kind of reflecting on what's happened over the past few months. I've been thinking about what's changed. One thing is, I really don't feel much like a medical student anymore. Occasionally I get into situations where I remember what being a student felt like, when I have no idea what I'm supposed to do. That's what being a medical student is all about, always with an undefined role. When that happens now, I remember how frustrating it was. I am more comfortable with making decisions now, but I don't think I'm ready to dictate those decisions to other people the way residents do. That's still frightening to me.

I'm staring to realize what I need to do to become a better doctor. I've got to become faster and more selective, be able to narrow things down quickly and home in on the diagnosis, because those are the things I'll need to be good at when I'm a resident.

So anyway, I guess I'm starting to become a master of internship, which is supposed to happen around now. I've become damn good at being a scut puppy, a data gatherer. I have a couple of tough months ahead: Infants' (pain and torture but with some good people); and a month in the Jonas Bronck OPD, which will be great but tough; and then my last month here, 6A. What a good-bye kiss!

Thursday, March 13, 1986, 1:30 A.M.

I got back from the West Bronx ER a little while ago. It was a typical West Bronx night. As soon as I walked in, Andy Ames signed out a child-abuse case to me. It took the usual form of no one understanding where the second-degree burn on the child's right leg came from. The social worker who called in the case had naturally gone home, and Andy was also gone, so that left me in charge. When the father came in angry and hostile, he couldn't find anyone but me to threaten. Everything was getting out of hand, and then the police showed up to start their investigation and that led to more havoc. Christ! Anyway, the BCW finally decided that since there was no obvious perpetrator—that is, no one had come forward and said, "Yes, I did it, I was the one who burned the baby," they let the kid go back home with the parents. I said, "Fine! Let him go home. What the hell do I care?" That's typical of the BCW! And what usually winds up happening is the kid'll show up next week or next month or next year dead. But what can you do? You can't fight the parents *and* the BCW. That's a little too much to take on.

The rest of the night was the usual. We had a bronchiolitic [**a child with inflammation of the bronchioles, the small airways leading from the larger bronchi to the lungs; children with bronchiolitis are usually under one year of age, and have respiratory symptoms that are very similar to those of asthmatics**] who probably has pneumonia [**since bronchiolitis is caused by a viral infection, it's not unusual that pneumonia, or inflammation of the lung itself, is often an accompaniment**] who bought himself a bed on 6A. I also saw this girl, a skinny seventeen-year-old who had hematuria [**blood in her urine**] and stabbing pain in her right lower quadrant. When I told her I had to do a pelvic exam, she refused. She said she'd allow it only if someone from Gynecology did it. I paged Gynecology three times and they didn't answer. The next thing I knew, the patient's uncle was calling from a phone booth on Jerome Avenue. He said the girl had got fed up with the whole thing and just walked out of the ER and he followed her down to Jerome. So he was calling very apologetically to say that she wouldn't come back. Right under our noses, she just walked. She was actively bleeding from somewhere; whether it was her vagina or her uterus, God only knows. But she up and left. Unbelievable! So I got on the phone with her and said, "Look, you know you're leaving against medical advice. I advise you to come back to the emergency room right away." She said, "No way! No fucking way!" So I said, "Promise me one thing: If you start to bleed profusely, you'll go see another doctor." She said, "Well, maybe." That was it. She just walked!

At about ten o'clock, the ER filled with exhaust fumes from the ambulances parked outside the emergency entrance. Exhaust fumes! That was great for the asthmatics. They thought they had come in to get treatment for their asthma; they wound up leaving in worse shape than they'd been in when they first got there! And the place was scorching hot for several hours; it must have been in the mid-eighties in there. God knows why! I felt very rundown and I had no appetite. I ate nearly nothing the whole night. I didn't want dinner. That whole child-abuse case was getting me down; it killed my appetite. But we finished at one, which isn't bad, and I came back home and listened to the messages on my machine. I ate some food and I'm listening to this music now and suddenly I'm on vacation. Tomorrow I'll be home! Strangely, I'm not that excited about it. I am excited about seeing Karen and my parents and everything, but I'm not excited about the idea of going home itself. It's funny, I think it's really starting to bother me that I'm going to be leaving the Bronx for good in a couple of months. I'm starting to feel that I've made some good friends here and I know I'll have to leave them and I'm already getting sad about it, three and a half months ahead of time. Isn't that terrible?

Amy

Wednesday, February 26, 1986

It's the last night of my last vacation of internship.
Tomorrow I start on 6A [at West Bronx]. I haven't worked
there before, but I've heard it's a real killer. And of course
I'm on tomorrow night. So I've gotten myself really
depressed.

These past two weeks have been very special to me, very
relaxing and calming and restful. This was the first time I've
been able to be a full-time mother, twenty-four hours a day,
seven days a week, without interference. Since I started my
internship, Larry and I have never been alone with Sarah for
such a long stretch of time. There's always been someone else
around. This was my first opportunity to get to know my
daughter. I did everything for her: I changed her diapers and
fed her her meals, I got to talk to her and to watch her go
through her normal activities without any interruptions. And
I actually managed to watch her take her first step! It happened about a week ago, while we were in Florida. She's been

cruising for a while now [cruising: **walking while holding on to a surface, usually a bed or a table**], but one morning last week she just let go and took three steps without holding on. It was great.

So I really got the chance to know what being a mother is about during this vacation. And I liked it. I liked it a lot. It sure is better than working in all these damned hospitals where nobody cares about anything except themselves. I really don't want to go back. I just don't want to go back to work tomorrow.

So what else can I say about the vacation? We stayed at a condominium in Fort Lauderdale. We went to the Miami Zoo, we went to the beach, we went out on day trips, we did a lot of things. I caught up on some sleep, and I had a lot of time to think about what's happened over the past few months and especially about what happened at the beginning of February. The more I think about it, the angrier I get. I really was taken advantage of! There was no need for the chiefs to do what they did to me. They definitely could have let me go home and found somebody else to cover the ER that night. It wouldn't have meant that much to them, but it sure meant a lot to me! I thought that going away, taking some time off, would make me ease up on this. But it didn't. I can't forgive them. And I can't forget it.

Well, I'm going to put Sarah to sleep now and then I'm going to try to relax a little. I'm really very tense about tomorrow.

Saturday, March 1, 1986

So far, 6A hasn't been as bad as I thought it would be. The census is low and it's a good thing because the chief residents are trying to screw me again. It may actually work out in my favor this time. I'm sure they won't be too happy about that!

What's happening is, there are usually four interns working on 6A. It's a big ward, there's the capacity to house fifty patients, so when it's busy, you really need to have four interns. But this month, we're one person short. That's because the fourth intern is supposed to be a psych rotator **[psychiatry residents have to work for four months on either internal medicine wards or, if they're interested in child psychiatry, on general pediatric wards during their internship year]**, and he's not going to be showing up. The reason he's not going to be showing up is that the people who run the psych program felt he was too "psychiatrically unstable" to do a rotation on a ward as stressful as 6A. So rather than having four people covering, we have only three people. It's too stressful for the psych rotator when there would have been four, but nobody's concerned about how stressful it's going to be for us now that there are only three. That's typical, typical! But I just might luck out because of the solution the chiefs have come up with.

What they're doing is this: Usually the interns are divided, two working with a senior resident, the other two working with a junior resident. Because we were short, the chiefs decided that the other two interns would work with the junior resident and I would work alone with the senior resident. The senior this month is Ben King, who is one of the best people in the program. He was the person who let me leave the morning after my last night on Adolescents' so I could catch the flight to Israel. So I'm very happy to be working with him. The junior resident is Dina Cohen, who's one of the worst people we have.

Because there's only one intern on our team and two on

the other, we started off the month with only one third of all the patients. And not only that, but on that first morning, Ben was smart enough to realize that most of the patients who were assigned to us didn't belong in the hospital in the first place. I started Thursday morning with seven patients. When we made rounds, Ben decided that three of them could be sent home right away, so I was down to four. I got only one admission Thursday night, and one of the other patients went home yesterday, so tomorrow I'll start with only four patients. That's not bad for 6A; that's not bad for anywhere. And none of them is what you'd call sick. Two are preemie growers. **[Six-A serves as an "overflow valve" for the neonatal intensive-care unit; when the unit gets crowded, preemies who have outgrown the problems of prematurity and only need to gain weight are transferred out to the ward. These babies frequently continue to have more problems than normal, healthy babies of the same age, however. Caring for a preemie who graduated from the unit is not just a baby-sitting service.]** One is a kid with AIDS who's here just because he's got no place to go. And one is a six-month-old with meningitis who's doing pretty well; he's just in the hospital to finish his two-week course of antibiotics. So, so far I don't have much to do. I'm not complaining about it. I know it won't stay like this for long.

Thursday, March 6, 1986

I was on last night with Dina Cohen. Jesus, what an airhead that woman is! She's completely incapable of making a single decision. She's totally incompetent. Yesterday afternoon, Margaret was signing out and she told me she had this adolescent girl who had come in the night before with abdominal pain and a positive urine pregnancy test. An emergency sonogram had been done that showed something around the right ovary. Ben was sitting next to me, and when he heard Margaret say all this, he got very upset because he knew the girl had to have an ectopic pregnancy [a pregnancy in which the gestational sac implants someplace other than in the wall of the uterus; it is dangerous because it can cause a massive hemorrhage]. Ben asked if Gynecology had come to see her, and Margaret said, "No, they haven't even been called yet." Ben just about blew his top! He ran over to Dina and asked her about it and she said, "Well, the ultrasound attending said it wasn't a conclusive study. He thought it could either be an abscess or a cyst or an ectopic—" Ben interrupted her and yelled, "It is an ectopic! You have to call Gynecology right now!" And Dina said, "Well, I'd rather be sure first. I think we should do a beta HCG [a blood test for pregnancy; more accurate than the usual urine pregnancy test], but you can't get it done until tomorrow morning. Can you imagine? You can't get a beta HCG done in this hospital after twelve o'clock—" Ben stopped her right there and said, "Dina, think a minute! You've got an adolescent with abdominal pain, a positive urine pregnancy test, and a finding consistent with an ectopic on ultrasound. You don't need a beta HCG. What you need is a gynecologist. They have to take her to the OR right now or she may bleed out before tomorrow morning!" Then Dina said, "Well, I thought we should do more tests—" And Ben said again, "You have to call Gynecology right now. Don't you understand?"

He finally convinced her to make the call. They came and saw her at about five o'clock and took her to the OR almost immediately. Of course, she had an ectopic. If Dina had waited and hemmed and hawed a while longer, that girl might have bled out right there on the ward!

Needless to say, I didn't feel very comfortable being alone with Dina for the rest of the night. The girl with the ectopic did okay. She stayed in the recovery room for a few hours and then came down to the floor at about midnight. And luckily, it was a quiet night; I only got one admission, and that was an asthmatic who didn't require any kind of expertise in his management. So I didn't ask Dina for any help all night long. I didn't even see her after midnight; she went off to sleep in her on-call room.

Sunday, March 9, 1986

It was seventy degrees today—really spring. We spent the day out on the grass in front of the apartment building. Sarah loved it. She got a chance to toddle around and see some of the other children. It was a really nice day.

Things on the ward are okay. I'm still getting along well with Ben. He's just great, the best resident I've ever worked with. He's got good judgment, he knows what's important and what isn't, and he's got a good sense of humor.

But the other team is having a lot of trouble. Because Ben knows when to send patients home and Dina doesn't, they now have almost all of the patients on the ward. And to make matters worse, the interns are finding it very stressful to work with Dina. Laura's doing okay; she's a very good intern and she doesn't have to rely on anybody for much help. But Margaret isn't as secure about herself and she's having a very hard time with Dina. On morning rounds on Friday, Margaret completely fell apart. She started crying, saying she couldn't go on. She refused to work up a patient. She wound up spending a few hours in the chief residents' office. She just needs more help than Dina can give her. I think there's also a lot of other things going on in her life now. But that's true of all of us, isn't it?

The chiefs decided that they had to get Margaret out of the hospital for a few days. How nice of them! She's stressed, so they give her the weekend off. My daughter gets the measles and is sicker than most of the patients who come to the emergency room but I still have to work! That's typical! Oh, what's the use? What's the use of talking about it and thinking about it over and over again?

Anyway, Margaret got to go home Friday afternoon and didn't have to come in for her call on Saturday. She's got a great medical student, Susan, who's running the service for her. Susan took Margaret's call on Saturday and did a good job. She did as well as a lot of the interns. I wish I had a student like that. I haven't had a good student all year. All my students have ever done is complain. They don't want to do scut, they don't want to run to the labs, all they want to do is stand around and be spoon-fed information twenty-four hours a day. Susan isn't like that; she's willing to work. A student like Susan can make internship a whole lot easier.

Actually, having a good student doesn't matter for me right now. I've still been having great luck on call. I've gotten only one or two admissions per night. I left yesterday with only three patients, and one of those is probably going to go home tomorrow. That'll leave me with only my AIDS patient, who's really just a social hold, and one of the preemie growers I picked up when I started. Neither of them requires any real work. I should be able to leave tomorrow right after work rounds. I'm planning to go home at ten o'clock in the morning! It's amazing!

Saturday, March 15, 1986

I just finished putting Sarah to sleep. This is my weekend off. It's been a strange week. Work's fine, there's no problem there. But Larry got called away on business on Monday. He's in Switzerland and he'll probably be gone all next week. It's been hard for me. I've had no relief in taking care of Sarah at night. Usually, when I'm post-call, Larry handles her 3:00 A.M. wake-up; usually I don't even hear her cry. But on Wednesday it was all me, and I was tired. I don't know how I got through it.

We got Marie to stay over on the nights I don't come home. She wasn't very happy about doing it and it's costing us an arm and a leg, but what else could I do? Someone has to be with Sarah; I can't take her to the hospital with me. I'll be happy when Larry gets back. His being away like this makes me realize how much I depend on him!

I haven't been feeling very well, either. I've been very tired. And I've lost my appetite. I think I've just made it to the point in the year where I'm simply exhausted all the time. I know a lot of the other interns have gotten to this point already; I'm surprised it's taken me this long to get here. And being tired sure isn't making it easier to take care of Sarah by myself!

My luck has been holding: I got two admissions last night; and when I was on this past Tuesday, I didn't get any. But that doesn't mean I've been getting a lot of sleep. The other team's been getting killed on their on-call nights. They've got a lot of patients, and some of them are really sick. So I still wind up staying up, doing scut on their patients all night long. More IVs fall out on 6A than anyplace else I've ever worked. I don't know what it is about that ward! Some of the other interns say the nurses actually pull the IVs out, but I don't believe it; to pull out an IV would mean the nurses were actually touching the patients. So far I haven't seen one come that close.

Friday, March 21, 1986

I know this is going to sound crazy. It doesn't make much sense, but it's true. I'm pregnant! I found out today. I think it's great, but I know everybody else is going to think I'm crazy.

I've been feeling lousy for a couple of weeks now. I've been run-down and a little nauseous all the time, but I just figured it was internship finally getting to me. And my period didn't come last week, but that's not so strange; it happens to me a lot. I saw Susannah in clinic last week and I was telling her how bad I was feeling and she said, "It sounds like you're pregnant. Is that possible?" I hadn't really even thought about it until then. I told her it was certainly possible, and she told me to send off a urine sample. I found out this afternoon that it was positive. Unbelievable!

Larry came home from Switzerland yesterday. I told him a little while ago. He thinks it's wonderful. We're both very excited. We'd talked about waiting until I was a junior resident before we tried again. I'm about six weeks now, which means I'm due sometime around next November. Sarah will be only about eighteen months then. That's closer than we had planned. And it means that for the last year and a half of my residency, I'll have two babies to worry about instead of one. But what the hell? One thing I've learned over the past few weeks is that we have to do what's best for us, and I think having this baby is the best thing for me, for Larry, and for Sarah.

I'm not going to tell anybody about this just yet. A lot of things can happen. I had a miscarriage in my first pregnancy, and that can certainly happen again. And anyway, the chiefs probably are not going to be exactly thrilled when they hear about this. But I don't care. That's their problem. I really don't care what they or anybody else thinks.

I just hope I can make it through next month in the neonatal intensive-care unit! If I continue to feel the way I do now, it isn't going to be easy.

Mark

Sunday, March 16, 1986

It's taken me a while to get back to normal, but here I am, having as much fun as I had during the first six months of this nightmare. Yes, even though I came that close to doing a triple gainer off the top of Jonas Bronck Hospital just two short weeks ago, life's now become a barrel of laughs again.

This last month has really been pretty disturbing. I mean, I always had this idea that I was immune to getting depressed or something. I really didn't think anything could get me down. I guess I just happened to stumble on the secret recipe for major depression: You take one garden-variety intern, deprive him of sleep for a couple of months, make him eat take-out pizza every meal during that time, and force him to take care of the sickest babies on the face of the earth. Mix well and let him marinate in his own juices for three weeks. Then you collect the pieces in a body bag and send them off to the morgue. I guess I managed to interrupt the process right before I made it to the final step.

I really can't take credit for saving myself. Carole really did it. She took care of me, and I'm really thankful to her. Our relationship had been going down the tubes over the past few months. And it's all been my fault; I mean, I've kind of had other things on my mind, like sleeping and eating, and I haven't been paying much attention to her. So things hadn't been great between us. I was starting to have some doubts that our relationship would make it through the year; of course, I was also having some doubts that *I* would make it through the year, so my concerns about the relationship were not exactly at the top of my list of things to worry about.

Anyway, over the past few weeks Carole has just about moved into my apartment. She's somehow figured out how to get rid of all the cockroaches. I have to admit, things are nicer without all the wildlife even though they had kind of become my pets. She's been here every night when I came home from work and she's been really understanding, listening to me complain about everything imaginable, from how much I hate my patients to the fact that the West Bronx coffee shop was closed down by the Health Department because of "unsanitary conditions." (The amazing thing is, that was the best coffee shop in the system. I guess mouse droppings and rat hairs really do make everything taste better.) I know I wouldn't have gotten back to normal, if you can possibly call what I am now normal, if it hadn't been for her being here when I needed her. Well, enough of this; it's starting to sound like a sermon or something.

So for the past couple of weeks I've been working in OPD on the west campus. It certainly has been a welcome relief compared with the neonatal eternal-care unit. I like the emergency room because it gives you the chance over a very short period of time to torture a large number of children who, if you're lucky, you'll never have to see again. That's a unique opportunity. It almost makes being an intern seem like fun. Not quite, but almost.

I was on call on Friday and it turned out to be Fascinoma Night in the West Bronx ER. Every patient who came in between the hours of 5:00 P.M. and 8:00 P.M. had some bizarre diagnosis or some record-breaking laboratory result. The very first kid I saw was this one-month-old whose mother said he had vomited every feeding since he was discharged from the nursery. Every feeding! Now, I immediately recognized that there was some sort of problem here. I mean, I may not be Sir William Osler [**a famous physician of the nineteenth century who was known for his legendary clinical acumen**], but I do know it's not normal to vomit every single feeding of your entire life. At first, I was a little skeptical about the story. It's a little hard to believe something like that, so I asked the mother why she had waited so long to bring the kid in. She said she hadn't waited long at all, that this was the fourth time she had been in an ER, and that no one seemed to want to do anything to find out what was wrong with the kid. Okay, so then I figured the woman had to be a fruitcake or something. I mean, any doctor seeing a baby who had vomited every single feeding of his entire life would get very concerned and do something definitive, wouldn't he?

I guess not, because when I saw the baby, it became pretty clear the mother had to be right. I couldn't believe it. He looked like a baby concentration-camp survivor. He looked worse than Hanson did on Infants', which is pretty damn bad! This kid was a pound and a half below his birth

weight, for God's sake! I did an exam and I didn't find anything. Then I sat and fed him for a few minutes. He seemed
to do fine right away, but about ten minutes into the feeding
he started to cry, and the next thing I knew there was baby
vomit all over my sneakers. Great! So I ran over and got the
attending and showed him my shoes and told him I thought
the kid had pyloric stenosis. [**This is a condition caused by
enlargement of the muscles at the junction between the
stomach and the first part of the intestine. Because of the
muscle enlargement, flow of partially digested food is
obstructed, and once the stomach fills, the feeding is
vomited. Pyloric stenosis is surgically repaired.**] He
refused to help me clean off my sneakers, but he did come see
the kid and we felt the abdomen, and sure enough, the kid
had an olive [**a mass in the abdomen overlying the site of
the stomach**].

But the pyloric stenosis isn't what made the kid so interesting. What made this kid a fascinoma was the fact that
because he was so sick, I sent off a blood gas to see how alka-
lotic he was. [**Because the infant with pyloric stenosis is
vomiting stomach contents that contain hydrochloric
acid, these children frequently manifest alkalosis, or lack
of a proper amount of acid in their blood.**] It turned out
he had a pH of 7.76, the highest recorded pH in the history
of the pediatric chemistry lab at West Bronx. I think as a
result of having my name on the lab slip as the doctor of
record, I'm supposed to get a commemorative plaque or
something. As a result of having the record-breaking pH, the
kid is getting a no-expenses-paid trip to the ward at West
Bronx for fluid and electrolyte therapy before the surgeons
take him to the OR tomorrow to fix his stomach.

And that kid was just the beginning. A little later we got

a call from one of the orthopedic attendings who told us that a patient of his was coming in. He told us, matter-of-factly, that the kid had been bitten by a horse. A horse! Where the hell is a kid from the Bronx going to find a horse to bite him in the middle of March? I mean, we're talking about the South Bronx here; this isn't the Kentucky Derby! So the story seemed a little peculiar to begin with. And then the kid showed up, and it got even stranger.

I didn't actually see the kid right away. I heard her first. The sound she made was very much like the sea lion tank at the Bronx Zoo around feeding time. And that was just her breathing! I walked over to see what was going on and the mother said, "Don't intubate her, she's got a problem with her trachea, she always sounds like this." It was at that point that I started to get a little suspicious. It turned out, this was the orthopod's patient. She was this five-year-old with some horrible disease called metachromatic leukodystrophy, a really rare metabolic disorder. She was followed by one of the neurologists. So I'm sure this is the only case in the history of recorded medicine of a kid with metachromatic leukodystrophy having been bitten by a horse. At least having been bitten by a horse in the South Bronx during the month of March.

I have to admit, the story kind of piqued my interest, so I asked the mother how the kid got bitten. She told me the girl was involved in this therapeutic horseback riding program and she had been out for a ride that afternoon. At the end of the ride, the girl usually feeds the horse a carrot. She did it this time but she forgot to pull her hand away after the carrot was gone, and the horse, not understanding the difference between carrot and hand, continued to nibble. So she got bitten. Right; that made all the sense in the world. I'm glad she cleared that up for me!

Luckily, the kid wasn't too bad off. The horse had broken the skin on the back of her right hand a little, but it didn't look like any bones were broken. The orthopod came in and we did some X rays, which were negative. He put a dressing on the hand and asked me what antibiotic I'd recommend to cover the bacteria from a horse bite. Since I've had so much experience with treating horse bites in the past, I decided maybe I should look it up. You know, in all the pediatric and infectious disease textbooks I could find, not one of them even listed horse bites in the index! Unbelievable! It's such a common complaint, I thought there'd be long chapters on it wherever I looked. Anyway, we decided just to treat her with broad-spectrum coverage and see her back in a couple of days.

I go to talk to this kid's mother while we were waiting for the X rays to be developed and she told me the kid had been completely normal for the first year and a half of life and then started to deteriorate. She's been going downhill ever since. It's really a horrible story. There's nothing anybody can do to help her. It's only a matter of time now. It's really sad.

Anyway, so that was Fascinoma Night in the ER. And we didn't even get out late, which is probably the biggest fascinoma. I like nights like that. Maybe "like" is too strong a word. I can tolerate nights like that. They don't make me want to jump off the roof after I'm done with them.

I had this weekend off but I'm on again tomorrow. I've got to get some beauty rest now. At this point, I'm at least six months behind.

Friday, March 21, 1986

Things have been pretty quiet, but I was on call last night and something did happen that I really want to get down on tape. At about eight last night this one-year-old came in with a fever. I called him in and started getting the history. He looked really familiar, but his name didn't ring a bell. He was brought in by this woman who was his foster mother who said he'd spent the first five months of his life in the hospital. She said that she wasn't sure what exactly had been wrong with him but that he had been really sick for the longest time and that all the doctors were sure that he was going to die. It was then that I realized that I was standing over Baby Hanson.

Hanson! I only have to say the word and I get nauseated and want to run to the bathroom to throw up. But he looked great. That puny, disgusting, horrible bag of piss-poor proto-plasm had grown into what looked like a fairly normal kid. He got taken away from his biological mother and placed with this foster mother in December, a couple of months after I had last seen him, when he was admitted to the ward at Jonas Bronck. The foster mother didn't know anything about the biological mother, so I don't know why he had finally been taken away from her. The foster mother had given him her last name, and that's why he was now Rodney Johnson.

It was amazing! He was sitting up, he could stand hold-ing on, he wasn't even that delayed. He could even say a few words, although he couldn't say "crump," which really should have been his first word. And he still didn't have a vein in him. I looked all over, just out of curiosity. Amazing! If you would have told me back in August that Baby Hanson could have grown up into this kid, I would have called a psych consultant for you. But there he was!

He only had an otitis [ear infection], and I sent him out of there with some amoxicillin. But I learned something from seeing him. I learned that no matter how horribly disgusting and wretched a baby is, there's always a chance he could grow up into a seminormal child. I never would have believed it.

I finish in the OPD next week and then I go on vacation. Carole and I have decided to go to Cancún. We were thinking about going back to that hotel in the Poconos we went to during my last vacation, but Carole decided against it. She thought I had been tortured enough for one lifetime over the past few months. I still think that maybe we should go. I mean, if I go someplace nice and actually have a good time, how am I going to be able to come back to the Bronx to finish the last couple of months of this wonderful experience? But who knows? Maybe Cancún will be hit by an earthquake or some other natural disaster, just to keep me in shape!

Bob

In 1981, three reports of an apparently new disease appeared in a single issue of the *New England Journal of Medicine*. The articles described a series of patients who had become sick over the previous few years with some serious and unique symptoms. The patients shared a great deal in common: Each had been in excellent health before the appearance of the illness; each had developed pneumonia caused by *Pneumocystis carinii*, a parasite that only rarely caused problems in otherwise healthy individuals; many had also developed unusual malignancies, such as lymphomas and Kaposi's sarcoma; and, in retrospect, all were gay men. These articles, which at the time appeared to be the result of chance coincidence, would have incredible implications. They signaled the beginning of the age of AIDS.

The story of acquired immunodeficiency syndrome in children began at our hospitals. In 1979, two unrelated children were referred to the pediatric immunology clinic at University Hospital with serious, recurrent bacterial infections, including pneumonia. These children presented a puzzling picture of immune deficiency not previously seen in the pediatric age group. By the time those first articles on AIDS appeared in the *New England Journal of Medicine*, five children had been identified with symptoms that were identical to those reported in the gay men. In addition to their recurrent infections and immunological abnormalities, these five kids shared one common factor: All had been born to women who were drug addicts. And these women were also becoming sick, developing symptoms very similar to those of their offspring.

The widespread acceptance of the fact that AIDS could occur in children did not occur until 1983. But whether accepted as fact or not, by 1983 it had become clear to everyone working in the Bronx that something terrible was happening.

Although pediatric AIDS started with a handful of cases, by the mid-1980s a full-fledged explosion had begun. The Centers for Disease Control in Atlanta estimate that by 1990 a total of three thousand children in the United States will become sick with AIDS. Scientists who have watched the epidemic develop believe this figure is an underestimate.

What this means to those of us working in the Bronx is that there are many infants and children who are or soon will become sick with AIDS and who will ultimately die because of it. At this moment, there are currently ten to twenty children with AIDS and the AIDS-related complex hospitalized in Jonas Bronck, Mount Scopus, University, and West Bronx hospitals. These children are in the hospital for one of two reasons. Some are critically ill; these patients have serious infections, cancer, and chronic lung disease. They fill beds in the ICUs for extended periods, draining resources and causing the staff who care for them severe emotional distress.

Other children with AIDS who are hospitalized are not sick, at least not initially. These kids live in our hospitals because they have no place to go. Their parents are drug addicts, many of whom have become sick themselves, and some have died. Grandparents and other family members have abandoned them; they've become pariahs because of their disease. Although some manage to escape from the hospital for some short period of time, most members of this group wind up living out their short lives knowing no home other than a steel crib in a three-bedded room at the back of 8 East or 8 West at Jonas Bronck Hospital, knowing no family other than the nurses, house officers, and medical students who provide their care.

But these hospitalized patients are only the tip of the iceberg. There are over a hundred sick children in our system currently being followed by the immunologists. Another hundred have already died. And these numbers are growing daily. Unless a cure is miraculously found, all these children will presumably die.

There's no question that AIDS has altered every aspect of modern medicine. It has radically changed residency training in virtually every specialty, including pediatrics. When I was an intern, there were few deaths on the pediatric wards. In fact, one of the reasons I chose to specialize in pediatrics was because children tended to recover from illnesses. But thanks to AIDS, all that's changed.

Amy, Mark, and Andy, as well as every other intern and resident in our program, have each been involved with at least one sick and dying AIDS patient. Over the past couple of years, about one child with AIDS has died every month. Occasionally the death seems almost like a blessing; these children are alone, with no loved ones; they are comatose, lingering on day after day in a vegetative state, with no hope of survival. But most of the time the death of a child, any child, is a tragic, deeply disturbing, and anxiety-provoking event for the house officers, nurses, and other staff members who care for the child and who stand by helplessly watching, unable to do anything to alter the course, as the child grows sicker and weaker until he or she ultimately dies.

But the inevitability of the death of the patient is only one factor that's changing the way house officers approach their charges with AIDS. The second and perhaps dominant force is tied to our current knowledge of the way in which the human immunodeficiency virus, the agent that causes AIDS, is transmitted. House officers know very well that if they stick themselves with a needle that has been in the vein of an HIV-infected individual, they can become infected. And becoming infected is equivalent to a death sentence.

When I was an intern, we drew blood, started IVs, even did mouth-to-mouth resuscitation without giving it a second thought. We knew of few risks and little harm that could come to us from stabbing ourselves with a needle or breathing in secretions from a patient who had had a respiratory arrest. Now it's mandatory that all house officers wear gloves whenever sticking a needle through the skin of any patient, regardless of whether the patient is thought to have AIDS or not. This increased use of gloves has caused a worldwide shortage of rubber. New types of gloves advertised as being resistant to HIV are being marketed. In the emergency rooms, nurses have been issued goggles to be worn over their eyes when around a patient who is bleeding profusely. Some residents don surgical gowns and masks just to enter a patient's room. And forget mouth-to-mouth resuscitation! What was once knee-jerk reflex is now something that house officers, with good reason, try to avoid at all costs.

Even though these precautions are being taken and everyone is being very careful, there is still a great deal of fear about AIDS within the ranks of our house staff. I went out for a couple of beers with Andy Baron one night early this month. He looked terrible: He has lost at least ten pounds since the start of internship, and he was barely able to keep his eyes open. It was pretty clear he was depressed. During our third beer, he let me in on why: He's convinced that he's infected with HIV. "I've stuck myself with so many needles, there's no way I don't have it," he explained.

I told him he shouldn't worry so much, that every intern and resident has stuck himself or herself multiple times over the past five years and so far nobody's tested positive for HIV. Andy replied that the key phrase there was "so far." He's sure that it may not be today and it may not be next month, but within ten years he and most of the rest of the interns in his group are going to wind up coming down with AIDS.

"How do people get AIDS?" he asked. "Drug addicts get it from contaminated needles that have been used by people who are infected with the virus, right? If we stick ourselves with needles that have been stuck into the veins of children who are infected, why shouldn't we get it? We're no different from drug addicts. We don't have any magical protection."

There's really no way to argue with his reasoning. I think

it's pretty safe to say that at this point in the year, most of the interns would agree with Andy. This fear of AIDS has definitely changed the way the members of the staff approach patients. And it's not something that will go away or change in the near future. AIDS is here, apparently to stay.

Andy

Saturday, March 29, 1986, 11:30 P.M.

I'm not going to talk for long because I've got to go to sleep; I'm on call again tomorrow. I'm on Infants' [**NW5— Infants' ward**]; I started a couple of days ago and of course I was on the first night. I got totally fucking killed. I was assigned six kids to start and then I got six admissions and a transfer that first night. It was like thirteen admissions, because I never had a chance to get to know anybody. In any case, it was terrible; I was up all night with a cross-covering resident who was really pretty mediocre; he didn't help organize things at all. Then the next morning was a nightmare; I couldn't present to save my life, it was like being a third-year student all over again. I was tired and they stole my scrubs and . . . it sucked. I felt disorganized, panicky, and I got chewed out by Alan Nathan for delaying giving an antibiotic to a patient. And there was so much scut I couldn't get anything done; I didn't have any progress notes written, nothing! I barely got my admission notes done. I finally got to sit down and write my progress notes at seven o'clock. I had two days' worth of notes to write! I was post-call and I had to try to make some sense out of what had happened over the past thirty-six hours! I didn't get out until after ten-thirty, and that was my post-call night! Ten-thirty at night! I was there for thirty-six hours without a wink of sleep, working my butt off the whole time. I'm still tired, and tomorrow I'm on call again. I think that was probably . . . that may have been the worst call I've had all year. What a fucking nightmare.

Anyway, I'm sure my senior resident, Eric Keyes, whom I like a lot, thinks I'm a complete idiot by now. He probably won't trust me for the rest of the month. First impressions are pretty important.

Tomorrow I'm on call with another idiotic cross-coverer. I won't mention names, but tomorrow I'm on with one of the worst, least-liked second-year residents in the program. What a pain in the butt! You know, they don't give a shit when they're cross-covering, because they're out of there the next morning. They don't have to face up to things; I do! I have to clean up the mess through the entire next day! And then my next call after that is on Wednesday, the day I have clinic. Then I'm on next Saturday. You know what that's like? It's like having four lousy calls in a row. It sucks! I hate it. I really hate this so much. If tomorrow's anything like yesterday and the day before, I don't know how I'm going to get through this month. It's absolutely torture.

Right after vacation to come back to this! I can't tell you! My vacation was pretty good. I'll talk about that some other time.

Monday, March 31, 1986, 8:30 P.M.

I'm post-call again. Not so angry this time and not so unbelievably tired. I had a really easy night, actually; I got only one hit. It was easy, but still I was running scut until midnight. The guy I was on with turned out to be a completely obnoxious blowhard who at least is pretty smart. He's a total zero as a human, though. Two calls and two total-zero cross-coverers. But now that I got these two over with I'm scheduled to be on with really, really good people during my next two calls. It's just too bad it's worked out in this order.

The nurses are great on Infants'; I really like them. But the place is a zoo; the private patients drive me up a fucking wall. They make me wonder what I'm doing going back to a privately run system. This lady today told me she didn't want me to draw her baby's blood. Jesus Christ! These parents are so uptight and nervous, they always want to come in and see the procedures being done. They can't accept the fact that they shouldn't be in the room. What do they think we're going to do, break their kid's arms? I have to think of nice ways to say, "No, you should wait out here, we'll be back when we're done, it's best for the child, and it's best for you, too. So don't come in!" Usually I don't get nervous when I'm doing procedures, but this lady today was making me crazy! And I couldn't get the blood. It was the first time that's happened to me in months!

I got these new medical students today. Brand-spanking-new students, never been on a ward before, I have this big, hulking guy named Ronald; he seems very nice. God knows; maybe he'll turn out to be a tyrant surgeon a few years down the road. He looks like one. He looks like he's going to be an orthopod. Anyway, I did my best to teach him stuff today, to get him over the jitters of being on the floor with real patients for the first time.

I've been hanging around the hospital too much. I stay too late. I was there until seven tonight. It's ridiculous! Finally Keyes said, "Get the hell out of here, you're just making work for yourself." He was right! What you're supposed to do is write your notes and get the hell out of there and let the person on call hassle with your patients. I think I'll do that tomorrow. I say I'll try, but I never can; I'm never able to get out before five.

I'm already in bed. It's eight-thirty and I'm already in bed, can you believe that? I'm going to sleep, I don't care. I'm always sleep-deprived; sleep's like going out of style for me. This job is so damn stupid! It's just stupid!

I spoke to Karen tonight for the first time in about a week, because of my stupid on-call schedule.

There I go; I fell asleep again. God! So, I spoke to Karen tonight. She's doing all right. We didn't talk for a long time. I miss her. I can't stay awake any longer.

Monday, April 7, 1986

For some reason, I've had all these revelations over the past week. At least they seemed like revelations at the time. Coming back to them now, they really seem like just a bunch of mundane thoughts. I seem to have them on the scut run between the chemistry elevator and the hematology elevator. I have no idea why, but over and over I get these things popping into my mind while I'm in the back corridor by the back entrance to the kitchen.

One of the things I realized was that, at this point in the year, I feel like I'm getting stupider, not smarter. I know it's not true, but I think maybe it has to do with the fact that the barn door has swung open to the world of knowledge. I guess I'm just realizing what you really need to know to be a decent resident. It's unbelievable; I just feel so stupid. And it doesn't matter whether I read or not; I don't remember anything an hour after I'm finished with it. But I've got to keep persevering. It's funny; I thought I was smart a couple of months ago. I'm not!

I also had this thought about nurses and how night nurses seem to be universally weak in all places except maybe the ICU, where they're still good. I don't know why this is. At night, there seems to be a certain stereotype: the middle-aged, fat, black nurse who's kind of disgusted and noncommunicative. And while she may not be all those things, the stereotype of being noncommunicative and disgusted seems to hold true. I don't know why, but from hospital to hospital, it seems to be the case. And it's kind of distressing because at night there's nobody else there, and sometimes you need to talk to somebody about a patient, and these nurses, they just don't want to talk about anything! Everything seems to be an effort when you ask them to do something.

Monday, April 14, 1986, 2:00 A.M.

It's 2:00 A.M. and I've woken up for some reason from my precall sleep. I really should be asleep. I have insomnia. I keep thinking terrible black thoughts because last night I wrote up the protocol for the M and M Conference [**Morbidity and Mortality Conference, a teaching conference run much like chief of service rounds in which a patient who has died is presented; the clinicians discuss the disease process, and the pathologists bring the autopsy report and describe what really happened**] on Emilio, my patient with AIDS who died when I was in the ICU in February. Yesterday I got Emilio's chart from the record room and I wrote up a summary of what happened to him over the weeks I took care of him. It really hurt to do it, to go through that chart again and to see that he was deathly ill the minute he arrived and never improved and that he finally, finally, by the grace of God, died. I remembered how he suffered and how his mother would come and sit by the bedside for hours. About a week before he died, she told me how at times when the Pavulon [**a paralyzing medication**] was wearing off, before he got his next dose, she would see tears forming in the corners of his eyes. She knew he was suffering terribly. But of course he couldn't cry out because he had a tube between his vocal cords and because he was more paralyzed than not. And we constantly would do horrible things to him in our effort to save him from certain death. So I wake tonight with these terrible black thoughts that I'm going to get AIDS, die the same death that poor Emilio died, having my lungs pumped with ventilator air every second, and my limbs poked with needles by young physicians in training, and my neck or groin poked by the fellow trying to put in a line, or my lungs needled and cut, while I hear the doctors saying crass and horrible things about my death and illness, making fun of my debilitated state, while I'm lying naked on a table and shitting on a blue chuck [**a pad made out of the same material as disposable diapers, which is placed under incontinent patients**], the way poor Emilio did, with no dignity. Just pain. How he must have hurt.

And I think about the Infants' ward I have to go to tomorrow on call and all the sick children I have to take care of there, two of whom are trying to die on me all the time. I don't feel up to taking care of these fragile little things. I'm tired of being abused by the system, of having my sleep taken away every third night, of the stress I'm put under and the illness I'm exposed to, and the pain I have to see and cannot heal. I'm tired of dealing with parents whose pain I can never completely understand because there's never enough time. The only time I have to try to understand what's happening to them is the time I take away from my own sleep. It's a constant battle. The doctors who do the best with their own lives, who get the most sleep and get out the earliest, are the ones who don't talk to the families, who don't play with the children, who don't thoughtfully consider things. But I'm not that way; I'm not efficient. I spend time with the families, I talk with them, and so I get sleep-deprived.

Tuesday, April 15, 1986, 9:00 P.M.

There's something I haven't talked about yet, something that's really hard about training that most people outside of medicine don't have to deal with, and that's the sense of loss of social skills that happens after you've been working all night. You then have to interact with people in a complex fashion. You have to go on rounds and talk to other members of the staff. I very often find that I have no idea how I'm coming off to anybody else. If people laugh, I can't tell if it's because I said something funny or if I've done something really dumb and embarrassing and that's the only reaction they can have. Am I offending anybody? Do I curse too much? Should I just fart and get it over with? A lot of times I just can't tell, I can't judge what people are saying to me: Are they being serious, are they making a joke?

It's not so bad with other residents. They understand, they can say, "Hey, he's been up all night, he's just post-call." But what about the parents of my patients? What the fuck do they think is going on? I might be acting really weird. Do they understand it's because I haven't slept in two days? They must think I'm just batty or something. And that's not good, because here they are, trusting me with their most precious thing in life, and I'm acting really flaky. It's an ill-defined concern of mine, but it really bothers me.

Today my student had to give a presentation of a patient. It didn't go too well; it was very rough, to say the least. I like Ron, he's a good guy. He reminds me of how I was as a student. Real nervous, disorganized, can't think on his feet, that's just how I was when I started out as a third-year. Shit, I still get like that sometimes. Anyway, after rounds, Mike Miller, who's our attending, came up to me and said, "Andy, I think you guys have to work on your student's presentation. It really isn't very good." And I told him I would, it was on the top of my list of priorities. So a little while later I sat down with Ron and we went over how to write up and H and P [history and physical exam] and how to present it on rounds. I can imagine what he was feeling: defensive, embarrassed, humiliated. It's one of those awful rites of passage. I don't know, I didn't get much sleep last night, I was really tired, and I wonder what he was thinking. I don't know if I was coming off as a hard-ass, if I was being condescending. I didn't mean to be; I kept saying to him over and over that you're not expected to know this, nobody ever teaches you this. I spent about twenty minutes talking to him about this, telling him the same stuff over and over again. He probably thought, You asshole, stop repeating yourself! I hope he learned something from it, and I'll tell you, the next time I'm on call, next Friday, he'll just have to take the admission and just go over it with me first. When I'm done with him, he'll sound like a master.

It's a beautiful, sunny spring day today. I came home and

I really wanted to sit out on the porch with some friend and drink some beers, but I didn't feel like calling anybody; I guess I really wanted just to be by myself. I'm too fucking tired to talk anymore.

Saturday, April 19, 1986, 11:30 P.M.

Last night one of my patients coded and died. It really hurts to go through the story again, but I suppose I'll try.

It was a little five-month-old with bad heart disease, doesn't matter what type, who had been admitted several times before for congestive heart failure. This time she was coming in to get a cardiac cath done so they could plan her surgery. When I first saw her, she was in some failure: She was puffing away and a little cyanotic [blue], but her mother said she always looked that way. And so I got her plugged in and talked to her attending and he also told me not to worry because this really was her baseline, so it wasn't necessary to start oxygen. The cath was scheduled for the next morning. In the evening, she spiked a temp; there was no obvious source, and it was only a low-grade fever. I figured maybe it was the start of a URI [upper respiratory infection; cold]. The resident who was covering looked at the kid and said she had an otitis [otitis media, an ear infection]. She didn't have an otitis, no more than I did. But the resident insisted, so we gave her some amoxicillin and some Tylenol and she defervesced. But she had spiked and it was the day before her cath.

She went for her cath bright and early yesterday morning and she came back at about ten. We were on attending rounds. I saw her for a moment, at about noon. She had fallen asleep in her mother's arms, and her mom asked me not to disturb her. I told her I'd come back and see her later, after she'd slept for a while. When I came back, she looked a little uncomfortable, but not bad. I got called away to do something else before I had a chance to finish my exam.

At two o'clock we were called to see her because the nurse had noticed she was looking worse. We went in: There she was, pale, tachypneic, with cold extremities. She looked clamped down and shocky. We had a devil of a time getting a line in. Before we did, we got a blood gas: It showed she was quite acidotic, with a pH of 7.21. But after we got the line in, we gave her a small amount of fluid and she seemed to become more comfortable. Her repeat blood gas was improved; she was less acidotic.

A little later we decided we should give her a little Lasix [a diuretic] so she wouldn't go into congestive failure or pulmonary edema. When I went to give it she looked comfortable, breathing at sixty instead of eighty. I spent the afternoon darting in and out of the room; basically she looked okay. We put her in 60 percent oxygen by headbox to help her out. Her attending also kept coming in and going out all afternoon. He was concerned that she had suffered some sort of ischemic event [damage to the heart muscle due to lack of oxygen], but he didn't know when. He told me I might have caused it by drawing blood and introducing an air embolus, something I'd never heard of before. That sounded like a really ridiculous idea. He said it might have been that or it might have been the cath, but he sort of kept stressing that I had done something.

Anyway, at about five o'clock he was there, and he chastised Eric. He told him how foolish and unobservant he had been. He told him that the child was in respiratory distress, grunting and flaring, and that he'd noticed it an hour and a half before, but that he didn't seem concerned. During the afternoon, the baby had spiked to 40.5°C [**almost 105°F**]. We were very worried, so we got a chest X ray, drew some blood, and started the baby on antibiotics. Her attending told us the fever was just a "dehydration fever." I saw him put his hand on Eric's shoulder and say condescendingly, "I've been in this business for a long time. I can tell you that's all it is." He didn't want us to start the antibiotics. But we did nonetheless.

At about six-thirty I was writing my sign-out, trying to get home; Eric was with Kelly Jacobs, the other intern, almost at the end of evening rounds. Eric remarked to Kelly that the baby had a "preterminal look." She had a heart rate of ninety, which is slow. Suddenly her heart stopped, right in front of their eyes, a witnessed cardiac arrest!

When I first heard the scream "Call a code!" I jumped up; I knew it was my baby. I ran into the room; they were starting to position her to start CPR. I turned around and helped the nurses haul the crash cart in. Eric intubated the baby and I took over managing the endotracheal tube while he ran the code. I started ventilating the baby while Kelly started sternal compressions. He was counting "One one-thousand, two one-thousand," up to five, and I forced a breath in every time he got to five. Meanwhile, the nurses had ripped open the crash cart and people began to fill the room from everywhere. And we began to code the baby.

We did everything we could. We pushed four rounds of meds. Jon, the chief resident, came and stuck a line in the baby's external jugular vein. We poured in fluid and kept pushing meds. But every time we stopped the CPR and looked up at the monitor, there was nothing. Flat-line. Finally we put her on an Isuprel drip. Even that didn't work. Then Eric tried intracardiac epi. And when that didn't work, after twenty-five minutes, they called the code and declared her dead.

At one point, well into the code, I remember looking up and seeing the mother, horror-stricken, with her hands to her mouth, bent at her hip like she had been punched in the stomach, screaming with horror. And then Jon had pulled the curtain so she couldn't see in. When we stopped the code, Eric pushed me out of the room and told me to tell the parents. He said, "You go first, go tell them." Just for a moment I stood in the baby's room terrified that I'd have to go through this experience again; I've already had to tell three sets of parents that their child had died. But this would have been the worst of them, because this was my patient, I had admitted her, and because this was a baby who wasn't supposed to die.

But I was spared giving the news this time. By the time we left the room, the mother already knew. Word got out very fast; one of the nurses had told the baby's grandmother, and the grandmother had told the mother.

We went out of the room as the nurses came in to clean up the mess. As we passed through the hall, there were shocked, terror-stricken looks on all the other parents' faces. Then we saw the mother. She was panicky and crazed. She wanted to run in and see the baby. We had to hold her back; we kept telling her that she shouldn't go yet, that she should wait until everything had been cleaned up. She was screaming that she had to see the baby and we couldn't keep her from her baby! But we told her again that she shouldn't see her baby now, with all the needles and the mess.

Someone found a wheelchair and we got her into it; with a lot of effort, we pushed her into the house staff lounge. We got the father in and we got everyone else to leave, and we told the parents exactly what had happened. That's when they began to cry. We told them we had done everything,

everything possible, and that nothing had worked; there was never a response. They just couldn't understand.

Then finally one of the medical students came in and said it was okay for them to come and see the baby. The mother darted out of the room and we followed behind her. We stood in the hallway and we called for the attending and we called for the priest and we called for the social worker. There were a lot of crying, hysterical relatives filling the hallway, filling the ward, and panic-stricken parents of the other children stood uncomfortably at the edge of the doorways, not knowing what to say or how to act. It seemed to go on forever.

The father didn't stay in the room long; he couldn't bring himself to look at the baby. The mother stayed. When she finally came out, we took her, and the father, back to the house-staff lounge and we sat and talked for a long time.

After a while I left the room. I had to try to finish my work so I could go home. It was hard to concentrate on my other patients, but somehow I did it. At some point, Jon came in and asked me if I would go back and sit with the parents while he and Eric went to attend to some other business. I went in. It was just the parents and me. They sat there, upset but now calm. They asked me, "What will happen to our baby now? Where will you put her?" I told them the baby would stay in the hospital until they had decided what they wanted to do. They asked me if I thought an autopsy should be done, and I said yes, I thought one should, so that we could find out exactly what had happened. But they shook their heads no.

A little later, Jon and Eric came back with the autopsy permission form. They urged the parents to consent to an autopsy; the parents said they would think it over.

Before I left for the day, I pulled Jon aside and began to cry. I couldn't stop; I cried for the baby and for all the other children I'd seen die. I told him that I'd had other patients who had died and that I was beginning to feel like a death cloud. We went and talked and he reassured me it wasn't my fault.

Then the family met with the priest and the social worker. Phone numbers were exchanged; I didn't give them mine, but I thought that someone else had given them my number. Now I worry that I've lost touch with the parents forever. I wish I could be available to them.

When the family left, Jon, Eric and I were standing in the house-staff lounge. Eric cursed about how terrible this all was and then, in a very serious and angry tone, he said, "This job sucks!" We sat there silently, morose and upset. But then Eric began to imitate and make fun of some of the attendings in the most merciless way. And pretty soon, we were all laughing, and it felt so good to laugh because it had seemed like forever since I'd last done it. But as I was sitting there laughing, this terrible sadness came over me; I started feeling guilty for laughing at such a serious time. Then I began to sense a horrible, black feeling coming over me.

I left after that. The baby's attending had never shown up. I was exhausted, so exhausted. It was a very bad night, a night during which I thought about quitting. And so I got up, and walked home.

I tried to call someone, just to talk about what had happened. All I kept getting was answering machines. So I tried to get drunk, but I could barely finish two beers because I was so tired. It's been over twenty-four hours since it happened, but all day today I've been feeling depressed and upset. And I feel guilty as hell about it, even though I've been told over and over again that it wasn't my fault.

Mike Miller had me over to his house. He told me over and over that the baby's death was not my fault. We talked for hours. I really think Mike cares. He told me that the whole thing had happened under the eyes of the attending and that I was not to feel responsible for the baby's death. And yet I do feel responsible. I feel terribly guilty and terribly sad. And I need to find some place for this inside of me so that it won't eat me up, so that I can live with it. This was definitely the worst death of all deaths this year.

Amy

Monday, March 31, 1986

I'm in the neonatal ICU at Jonas Bronck now. Most of my patients shouldn't even be alive. They're so small and sick, I can't understand how any of these things are going to grow up to be anything like a normal child. Most of them don't even seem human. It's such a contrast for me. I spend the days running around doing all this worthless work on these premature babies and then I go home and see Sarah. Sarah's a real person now, she's walking around and talking. It's great to watch her and to be part of the process, but then I go back to work the next morning and I realize that no matter what we do or how hard we work, it'll never be possible for most of my patients to do any of the things my baby can do. I feel very sorry for them but I also feel sorry for myself, because I'm being forced to take care of them and I don't want to have to take care of them. I stand back sometimes and realize that we're not really doing anybody any good. But that's not for me to say. I'm supposed to do what the attending tells me.

And things are even worse this month because I'm feeling so bad. I'm nauseous all the time now and I'm always completely exhausted. I get to a point around three o'clock every afternoon where I have to lie down for a few minutes. Trying to get a half hour or so off in the middle of the afternoon in the NICU is not the easiest thing to do when all your patients are in critical condition and dying and there's still a ton of scut that has to get done. So far this pregnancy is a lot worse than my pregnancy with Sarah. I'm feeling a lot sicker. It's probably because I started out this one so chronically exhausted.

I still haven't told anyone in the program except Bob Marion that I'm pregnant, so I have to cover up how I'm feeling and make excuses about why I need to go off and be by myself for a while. I don't think anyone's caught on yet, but I'm sure it's only a matter of time.

I've got a total of eight patients. Four of them started out weighing less than eight hundred grams [**one pound, twelve ounces**]. Two of these are relatively new preemies and are extremely sick. Both are on ventilators; one has NEC [**necrotizing enterocolitis, a serious infection of the intestine**] and sepsis and all sorts of other problems. I spend most of my time trying to keep these two from dying. My other two preemies have been around longer; one's about a month and a half old and the other's nearly four months old. I remember the four-month-old from the last time I worked in the unit; he was very sick back then. Now he and my one-and-a-half-month-old are stable but they're both really damaged. They both have grade IV IVH [**intraventricular hemorrhage, a hemorrhage into the ventrical of the brain; IVH is** graded from I to IV, with IV being the most severe], and one has bad hydrocephalus. The four-month-old has been abandoned by his mother. The mother hasn't been around in over a month. I don't know if I can say I blame her. I'm not really sure what I would do if I were in a situation like that. God forbid! But the nurses are really attached to this kid. That seems to be something that happens a lot when babies spend so many months in the unit.

I've got another patient, a six-month-old, who was born with congenital hydrocephalus. He's been in the nursery his whole life! His head is enormous; it looks like a basketball. He's had five shunts placed, but none of them seems to have done any good. All he can do is suck, breathe, and keep his heart beating, so he won't die for a long time, but he also isn't going to be leaving this unit. It's really sad. His mother visits him every day. She's kind of a pain; she always asks about his head circumference and tries to find out what his last CT scan showed. I feel sorry for her. She's really devoted to him. I don't think she has much of a life outside of here.

Then I've got this whole assortment of crack babies who live in the ICU. We had some in the well-baby nursery when I was working out there but they were mostly just social problems. These babies in the unit are all very sick. They all started out addicted and SGA [small for gestational age]; a couple of them have had convulsions. And all of them have either been abandoned or have been taken away from their mothers by the BCW. They're all very pathetic. They start out sick, and even when they get well, they have no place to go.

So that's my service so far. I don't understand how I'm going to have enough motivation to get up every morning and go to work. I'd rather just stay home and lie in bed all day than go to that horrible ICU.

Saturday, April 5, 1986

One of my patients died last night. Dying was definitely the best thing that could have happened to this baby; he had almost no chance of survival. But dealing with the mother was very hard.

It was a baby who had been born two days ago with severe malformations. At first we weren't sure what the baby had, but we knew it was something really bad. She was very abnormal-looking; she had a loud heart murmur; and she didn't respond to anything, including pain. Bob Marion came to see her and said the baby had trisomy 13. [**This is a disorder caused by an extra copy of chromosome thirteen. Children with trisomy 13 have so many malformations, both internally and externally, that almost all die within the first few months of life.**] He told us we shouldn't do anything heroic to try to prolong life, so we didn't, we just let the baby be and kept her comfortable. It took her over twenty-four hours to die; she waited until everybody else had gone home, so it was only me and a senior resident who was cross-covering.

The baby's heart just stopped beating. One of the nurses called me over and told me she was dead. I listened and didn't hear any heart sounds. At least she chose a reasonable time to die; it was about eight-thirty. At least the nurses didn't have to wake me up at four in the morning to declare her.

The mother was out on the postpartum ward and I went out to her room to tell her. She was really upset. Bob Marion and Ed Norris, our attending, had talked to her a few times

since the baby had been born, so she knew what was wrong
and that it was only a matter of time, but still she got very
upset when I told her. It was really sad. She's thirty-nine years
old and this was her first pregnancy. I sat with her for about a
half hour, trying to calm her down. She told me she really
wanted this baby, that she had had a lot of trouble getting
pregnant and had gotten to the point where she didn't think
she'd ever be able to conceive. And then, just when she had
just about given up hope, she became pregnant with this
baby, who turned out to have trisomy 13. I had absolutely no
idea what to say to her. What can you say at a time like that? I
don't even want to think about it, especially not now. So that
wasn't the easiest thing in the world to deal with.

The rest of the night wasn't bad, though. I even got a few
hours of sleep. Most of the preemies behaved themselves and
didn't do anything stupid while I was on. Still, all night, I
kept thinking about that mother. And so even though it
should have been an easy night, it wasn't.

Friday, April 11, 1986

Yesterday was the official opening of the neonatal
intensive-care unit on 7 South. The unit's actually been open
for months now, but that's beside the point. They had this
big celebration, and Mayor Koch, a group of other officials
from the city, and all these reporters were here. Ed, our
attending, who's the director of the nursery, took the mayor
on a tour and showed him some of the six-hundred-
grammers. The mayor was, to say the least, a little put off by
the appearance of some of our patients. He didn't volunteer
to kiss any of them, the way most politicians kiss babies.

I was on last night, and I had a terrible night. At about
1:00 A.M., we got a call from the DR that they had this

woman who had just walked in off the street who was ready to deliver and had had no prenatal care. They weren't even sure when her LMP [**last menstrual period**] was, but they thought the baby might be about twenty-six weeks. Luckily, I was on with Enid Bolger, who's the senior in the unit this month. Enid also happens to be very good. We went running down to the delivery room just as the baby was coming out. It was tiny! I was positive it was too small to survive. Enid thought so, too, so we just gave it some oxygen and didn't do much else. It was only about eleven and a half inches long, so it was probably about twenty-three weeks [**about three weeks too early for the baby to live independently outside the uterus**]. But the baby came out with a heart rate, so we had to take it back to the ICU and wait for it to die.

That was pretty bad, but it could have been worse. Enid was very good about it; she didn't go crazy, like some of the other residents do. Some people would have done everything: intubated it, put it on a respirator, started IVs. But it wouldn't have accomplished anything. That baby never had a chance. It weighed only 520 grams [**about one pound, two ounces**]. So it was good to have someone like Enid in charge; I agreed with her completely in how this baby was managed. When the heart finally stopped beating, Enid and I went down and talked to the mother. She didn't seem too upset. I don't think she had really thought of it as a baby yet. I'm not even sure if she had realized she was pregnant before that day. She has three children at home. I don't know, I'm fairly sure it was for the best.

But I didn't get much sleep, and I've now had two deaths in one week. That's a lot, even if you're expecting the baby to die. I hope my luck changes.

I got into a fight with a mother last night. It was the

mother of my patient Moreno, the six-month-old with hydrocephalus. I was in the middle of doing my evening scut and one of the nurses told me that Mrs. Moreno wanted to talk to me. I told the nurse I was busy just then and I'd try to stop by later. It was a very busy night and I completely forgot to go back to find the mother, and about an hour and a half later the woman came up to me and said, "Didn't the nurse tell you I wanted to talk to you?" I very nicely explained to her that I had a lot of work to do and I couldn't talk just then. She then got nasty, accused me of neglecting her child because I didn't think he was as important as the other babies, and said she was going to go right down to the patient advocate's office to make a formal complaint against me. I felt so lousy and I was so fed up with everything that I just told her, "Go ahead! Go and complain! What can they do, fire me? Let them fire me! I'd be glad if they fired me!" And I went on like that for a while. The woman didn't say another word, she just walked off the unit. I guess she went and complained. I haven't heard anything about it, but I'm sure I will. The nurses told me not to worry, though, because since the baby's been in the unit, she's complained about four other doctors. Still, I don't think I should have been so nasty to her. But what can I do?

Thursday, April 17, 1986

I don't think I want to work in that unit anymore. Too many bad things have happened in that nursery. I don't like taking care of those things. I just want to stay home with Sarah.

I had another death last night. That makes three on my time in three weeks. This last one was the worst of all of them because it was my four-month-old preemie, the one the nurses were really attached to. He had been doing poorly all along, and last night at about two o'clock, one of the nurses went in to check him and found him dead. Just like that. He was cold already. There wasn't anything we could do for him at that point. The nurses were really upset about it; some of them cried. I've never seen a nurse cry before for a patient who had died. At least I didn't have to go through the stress of talking to the mother, but spending the rest of the night with the nurses who were in a rotten mood might have been even worse.

Mrs. Moreno did complain about me to the patient advocate. I got called down there on Monday, and one of the administrators asked me to explain what happened. I told him the whole story and he told me that this woman was very angry about what had happened and what was happening to her baby, and she was blaming the entire staff for everything. He said it was a bad situation but they were trying to keep her calm and that for the rest of my time in the ICU I should try to be nice to her and put up with all her craziness. He didn't fire me. I was hoping I was going to get fired or at least suspended for a week or two. No such luck.

There's only one more week to go, so it looks like I'm going to make it through the month. I go to the OPD next, and that should be easier. And I think I'm starting to feel a little better. I didn't feel like I had to vomit once all yesterday and today. So maybe this part of the pregnancy is coming to an end. I know that'll make things easier.

Mark

Tuesday, April 15, 1986

Well, I'm back from another wonderful vacation. I don't know what it is about me and vacations, but no matter how hard we try or how much in advance we plan, things always seem to turn out as if we were characters in one of those low-budget disaster movies.

This time Carole and I went to this beautiful hotel right on the beach in Cancún. It was a gorgeous place: great rooms, delicious food, and an amazing view. Everything would have been perfect, absolutely perfect, if two weeks ago hadn't turned out to be Mexican monsoon season. I didn't even know they had monsoons in Mexico, but I could have sworn that's what it was. We got off the plane at the airport in the pouring rain, and the rain continued the entire time we were there! An entire week of looking out the window at rain falling on a beautiful beach. Very exciting! And of course this place didn't have any indoor activities. Why should they have indoor activities? It never rains in Cancún, the weather's always perfect, isn't it? Sure it is, it's always perfect, except, of course, when I go there on vacation!

Everyone I've met since I got back to the Bronx has mentioned what a nice tan I got. Nice tan? There's no way I could have gotten a tan! There wasn't enough sun to get tan, the sun never came out, not for the entire eight days we were there! The only way my skin could have turned color is if maybe I started to rust. That must be what it is, I went away on vacation and got a nice rust!

Well, at least I got some rest. And I did get to spend a lot of time alone with Carole. That was very nice; we had a good time together. And I even got to read a book while we were away. An actual novel, not a pediatric textbook. It's the first nonmedical book I've read all year. It was getting to the point where I was starting to think I wasn't allowed to read any sentence unless it had at least one six-syllable word that had a Latin root in it. Reading this novel was tough at first. I'm so conditioned by reading medical stuff that I kept falling asleep after reading one paragraph. But I finally managed to get through it. I guess there's still hope for me.

Well, so much for my fond memories of our trip. It's now the middle of April, and the big news is there's only a little more than two months of this misery left. Yes, winter's over, the snow's all melted, the leaves are starting to appear on the trees, the addicts are starting to hang out on the street corners again, and the cockroaches are mating. I guess the cockroaches are always mating, but I swear, I got back to my apartment after flying back from Cancún and there were at least four times the usual number of roaches hanging around. We had managed to cut down the number remarkably for a

while but now they're back in full force. It was an awe-
inspiring sight, opening the front door, happy to be home
after narrowly escaping being killed in the floods, and looking
inside to see that my apartment had turned into Cockroach
Heaven! I've been fighting the war ever since, but it looks as
if they've definitely established a foothold. It's time to start
thinking about moving out of the Bronx! I think Carole may
ultimately win this battle. She's wanted me to get out of here
for months.

I'm spending the last two weeks of the month in the
OPD at Jonas Bronck. I've been on call two nights so far and
it really hasn't been bad. We've left the ER at the stroke of
midnight both nights, which is kind of amazing. We seem to
be between seasons right now. Respiratory infection season
has ended, and diarrhea and dehydration season hasn't
started yet. At least that's what all the attendings keep saying
to explain why it's so quiet. They talk about it like these diseases are sports. You know, in a couple of weeks I expect the
chairman of the department to come down to the ER and
throw out the ceremonial first diarrhea and dehydration
patient to open the season officially. Well, whatever's causing
it to be quiet, I'm not complaining. I just hope it keeps up.

I'm assigned to neurology clinic on Friday mornings.
What a lot of fun that is! The clinic is held on the fifth floor of
Jonas Bronck, and you get there at about nine o'clock and
the waiting area is already filled with what looks like hundreds
and hundreds of kids seizing and shaking and yelling at the
top of their lungs. Every kid has a chart that contains at least
five thousand pages, and you have to read the entire chart to
figure out what the hell is going on. And because there are so
many patients scheduled, you have to move really fast or else
you wind up staying all afternoon. This is the first time I've

been assigned to neuro clinic this year, and I'm glad I have only a couple of sessions there before the month ends. If I had to imagine what hell might be like, I don't think I could come up with anything worse than being permanently assigned to neuro clinic. Get the feeling that I'm not going to be a neurologist when I grow up?

Over the past few weeks, I've started thinking about what I'm going to do when I'm finished with this internship and residency. The more I think about it, the worse the headache I wind up getting. I can see the advantage of specializing in something, but I also can see the advantage of not specializing in anything. I could go right into practice after I'm done, or I could do a fellowship. So far the only thing I've definitely decided is to put off making a decision about this for as long as I can. It's really very early. I don't even have to start panicking until about next year at this time. I've got all the time in the world.

I'm also getting a little worried about July. In July I'm going to be magically transformed into a junior resident. I like the idea of not being an intern anymore, but I'm not so sure I like the idea of being a resident. I mean, residents are people the interns turn to when they have questions and concerns. Residents are figures of authority. For some reason, I can't seem to imagine myself as an authority figure. I can't imagine giving interns advice. I wouldn't trust advice I gave to myself. Of course, that sounds kind of ridiculous.

Bob

During the course of training, each doctor develops his or her own individual style of dealing with the family of a patient who has died. The evolution of this style occurs mostly through trial and error; with enough experience and having made enough mistakes, you gradually develop a method that makes both you and the family comfortable. There's no way this can be taught in a classroom or through reading books or articles.

Some doctors find that they feel most comfortable sitting down and having an open and frank discussion with the family, explaining to them in an honest and supportive way the events that led up to the death of the patient. This method is used most often by physicians who have had a lot of experience and who have a great deal of confidence in their skills. Young house officers have difficulty being very frank when discussing the death of a child with parents. Often there are many questions in the minds of the intern and resident about what actually led to the patient's death; they worry that they might have missed something important that could have saved the child's life, or that some task for which they were responsible was overlooked and contributed in some way to the patient's death. So interns usually don't feel comfortable having long discussions with the parents of a child who has died.

Some doctors overstep the traditional role of the physician and cry along with the parents of a child who has just died. This isn't necessarily a bad thing; the parents often appreciate the fact that their grief is shared by others who knew and cared a great deal for their child.

This style certainly described me as a house officer. While on the oncology ward in May of my internship, I cared for a twelve-year-old boy with leukemia. Tom's disease had been diagnosed the previous September, and coincidentally I had cared for him during that admission as well. Now we were both back on the ward, and on the first morning of the month, Tom was once again assigned to me.

It was a shock to see him after all that time. Back in September, he had been a strapping, healthy-looking young adolescent; by May he had been reduced to a wasted, comatose vegetable, unable to speak or eat or react to any outside stimulus. His mother, with whom I had become friendly during his first admission, stayed with the boy constantly during the time he spent in the hospital, guarding over him for what proved to be the remainder of his life.

I was on call the night Tom finally died. His vital signs had become very irregular during the evening; his nurse had called to tell me this news, and I had left what I was doing to come. We stood silently over him, his mother at the foot of the bed, the nurse to the left, and me on Tom's right, and we waited for his breathing to stop. That finally happened about an hour and a half after I had first entered the room. I was the one to declare him dead.

I had spent all that time in Tom's room not because I was his doctor; there was nothing I had learned in medical school or during my internship that could in any way have altered the course of events. I knew that, and Tom's mother knew that. I had stayed in his room because I was a friend; I had known him and his mother for nine of the most difficult months of their lives, and I was with them at the end out of respect for that friendship. I think my being there meant something to Tom's mother; I know it meant an awful lot to me.

But some doctors find that they can't deal with death at all. They equate death with failure, and they have trouble dealing with and accepting their own failures, and they have trouble dealing with and accepting their own failures. Once the patient dies, these physicians simply wash their hands of the whole affair. They leave the counseling of the family, the "mopping up after," to others.

In the case of Andy Baron's little patient, it seems as if the child's attending fell into the latter of these three groups. Unfortunately, the job of talking to the parents fell to Andy and the other house officers who happened to be around. At this stage in his training, Andy is looking to people such as that attending to guide him through this process. I'm almost positive Andy did a good job with these parents; I know him well enough at this stage to understand how sensitive and sympathetic he can be. But still, the attending's absence at this critical time must have been very difficult for the house staff as well as for the parents, who were looking for answers that couldn't possibly have been given by anyone other than the attending.

I know it must have been difficult for the parents, because

I've been on their side of the fence. Last year, my wife delivered a stillborn baby. Beth had started having what she thought were labor pains one evening about two weeks before her due date. We had gone to the hospital with all our stuff, figuring our baby was about to be born. When we got to the labor floor at University Hospital, a nurse listened to Beth's abdomen with a fetoscope, a special stethoscope designed to amplify the fetal heartbeat. She couldn't hear a thing. Without a word to either of us, she left the room, and about five minutes later, a resident appeared at the door, pushing an ultrasound machine in front of her. She introduced herself and told us that she was going to do a sonogram. And without saying anything else, she went about her work.

I've watched over five hundred fetal sonograms. I have a pretty good idea of what's occurring on the ultrasound machine's screen. And while watching the scan that the resident did, I could make out our daughter's head and her chest, her abdomen and her limbs, but I could not see her heart beating. After a few minutes of searching, the resident picked up the ultrasound machine's transducer, turned off the power, and said she'd be back in a few minutes.

"I didn't see the fetal heartbeat," I said.

"I didn't either," she said quietly, startled at my statement. "I'm going to call your attending."

Beth and I sat in the room crying; no one came to explain what was going on or what would happen next. Finally, after about a half hour, Beth's doctor appeared at the door. He told us that it seemed as if the baby had died. He talked with us, answered our questions, and told us what the most appropriate management plan was. Since the labor pains that had begun this whole episode a few hours before had completely ceased, Beth agreed to wait until natural labor resumed, an occurrence her doctor assured her would take place within the next week or so. And so, devastated, we prepared to go home.

After leaving the labor room, we approached the nurses' station. The nurses and the resident who had done the ultrasound were having an animated conversation, laughing and apparently enjoying themselves, but as we approached, they became silent. I was used to this; I had been part of this kind of behavior, especially during my time in the neonatal ICU. But now I was experiencing it in a different way; Beth and I were the opposition now, and this behavior made our grief just a little bit worse.

I've learned something from this experience, and accordingly I've altered the method I use when talking with the families of children who have died. This isn't an experience I would recommend, but it did help me understand a little more about what goes through the mind of a parent whose child has died.

Andy

Wednesday, May 7, 1986, 7:00 P.M.

I finally finished Infants'. It was a horrible, depressing month. About a week after the baby with heart disease died, I had another patient who got very sick and had to go up to ICU. I intubated him myself. He deteriorated so fast, he almost died on the ward before we could get him up to the unit.

Now I'm in the OPD, and all the details are starting to blur. I suppose if I spoke into this thing religiously every day, I could tell you endless story after story about all the kids and their various problems. But what does it matter? It's all just a horrible blur, one after the next, made up of all these poor, sick kids.

Thursday, May 8, 1986, 10:00 P.M.

It's become really hard to continue keeping this diary. Over the past few months I've lost touch with my inner self;

I'm not sure completely why that's happening, but I think it's because I'm defending myself against all the bad feelings I've had about being an intern. It relates to a lot of different issues having to do with the general feeling of being abused and mistreated, and the fatigue and the sleep deprivation, and the death and the morbidity of my patients. So certainly that's one reason I haven't been talking. I'm out of touch with myself, and it's hard to know what exactly to talk about. The other thing is, the thrill and excitement and novelty are gone, and they've been replaced by a more realistic perception of what I think medicine is. And for some reason, there's something in me that doesn't want to relate all those stories about all the various patients. Talking about it makes me feel like I'm back at work, and I hate even to think about being at work.

Saturday, May 10, 1986, 4:10 A.M.

I just spent the last six hours in the Jonas Bronck ER working on a fucking child-abuse case. I really hate them; I hate them more than anything else in this job. I think I've seen enough child abuse for an entire lifetime. I don't want to see any more, thank you. They never go well, they're always difficult.

This one, I just pulled the chart from the box, I didn't even read the triage note, and I called the kid in. She had a bandage on her forehead. Oh, great, I thought. A laceration. I asked what happened and the mother gave me this story that the girl was lying on the floor and playing and she bumped her head and cut herself on the hinge of her glasses or something weird like that and cut her forehead. I asked her to go over that again and the mother gave me basically the same story. So I took the glasses off the kid's face and I tried to find a way to make the hinge hit up against the fore-head. I couldn't do it. The frames were plastic and they were totally intact. I thought, No way! No way the kid could have done this!

So I decided I'd better do a complete examination. I got her undressed, and lo and behold, she had big contusions across her back and across her upper right thigh. I just thought, Oh, fuck! You get a feeling down in the pit of your stomach when you finally figure out what you're dealing with, and I got it at that moment.

Then I examined her vagina, and it looked kind of red and smelled bad, and I thought, Oh fuck! again. To make a long story short, I reported the kid to the BCW and the cops as a suspected physical and sexual abuse case. And I had to fill out only about a thousand forms among the chart, docu-menting the living shit out of it, the BCW 2221 form, and the rape evidence kit [**documents and materials that will be needed when the case goes to court**].

The whole thing was horrible. The parents were crazed; at one point they tried to take the kid out. They started to dress her and said they were going to take her to Washington Hospital [**a municipal hospital in the South Bronx**]. Give me a break! I called security at that point. Once I called secu-rity, that was it, they knew the jig was up. They knew they had been caught. Oh, man! It was horrible. I hate it. After I called security, I was shaking and nervous for a while because it's such a bad thing to have to deal with. I don't want to help take kids away from their parents! Kids don't want to be taken away from their parents; they love them even if they are horrible! So even though the parents have done something terribly wrong, I'm the one who feels like he's committing the crime.

Anyway, it takes so long to do everything, God knows what the kid's disposition will be. I don't know what to say. I hate it, I hate child abuse so much, I wish it never existed.

Wednesday, May 21, 1986, 9:30 P.M.

It's been a typical wild month in the Jonas Bronck ER. I'm getting out of this month exactly what I wanted: I'm learning how to manage trauma, and I'm learning how to see multiple patients in a short period of time. I'm a lot better at it than I was; I'm still not able to be as accurate as I'd like to be, but I can see some improvement every day. I can be fast when things aren't too complicated; I still haven't gotten good at seeing a complicated patient and a couple of uncomplicated patients at the same time. But I have another week in the ER and maybe I can get a handle on that.

I can't remember anymore what I've talked about and what I haven't talked about. I don't know, there are so many stories, so many stories of frightened mothers and frightened children, sick children, and I don't know why, I just don't want to talk about any of it anymore. I've had some bad nights, I've had some good nights. I'm sorry . . . I'm sorry this is deteriorating. But the year's almost over, it's just another five weeks or so and I'll be moving back to Boston. I really need to start making arrangements. I haven't done that yet. I'll have to take a day off from work to get that squared away.

Everybody seems to be calling in sick all the time now. Except me. There's one intern in particular who's always calling in sick, or coming late to clinic. I'm thinking that maybe I'll fucking call in sick one morning and get everything arranged for the move. But I think maybe this is another fantasy of mine. I haven't missed a day of work yet this year and

I probably won't start changing and calling in sick with so little time left. It pisses me off a lot when other people call in sick. It's totally irresponsible and everybody always winds up having to work a little harder to make up for the person who calls in sick, and that's not fair.

My clinic's going fine, and I have a couple of specialty clinics including renal, which I think I like a lot. I think I could actually do renal. I'm not sure yet, I'll have to try it again when I'm in Boston, but there are a lot of good things about it: It's interesting, it isn't a lot of hard work, and the people seem nice. I don't know, it's something I might be able to be content with for the rest of my life.

People have been saying a lot of nice things about me over the past few weeks. They tell me how much they're going to miss me and that I've added a lot to the program. A couple of the attendings have said that I'd make a good chief resident. That's all very nice and very flattering; part of me likes that fantasy of staying here and being asked to be chief, even though I know it's just a fantasy, and part of me now is very slowly, very slowly recognizing that I'm actually going to be leaving soon. I haven't started thinking of myself as being a resident in Boston next year; I don't have an emotional attachment to that program yet. I can see myself as a junior resident here much better than I can see myself as a junior resident there. I really wonder if I'll be ready for the demands of that place.

Next week I start my last month on 6A. My last month! My God! You know, this is going to sound tacky and very clichéd, but the year really has gone by fast. Two hundred ripped-off nights!

The chief residents' beeper party was today, and I missed it. I'd been looking forward to it for months. I even went out this morning and bought a blueberry pie to bring. Then I got stuck in the ER with a fifteen-year-old who got hit by a car and was dragged twenty yards. He was a mess; he had a basilar skull fracture, a hemotympanum [**blood behind his tympanic membrane, a sign of skull fracture**], blood in his urine, a laceration over the eye. I was fuckin' stuck with him and I missed the party but I learned a little about handling multiple trauma. I wanted to go so badly, I really was pissed off. One of the highlights of the year, and I missed it. Too bad. There'll be new chiefs the day after tomorrow. New chiefs: No more calling Jon, no more calling Claire, no more calling Arlene, no more calling Eric. I wanted to thank them, I wanted to thank them all, and now I don't know if I'll get the chance. I'll miss them, and I'll remember them. They were really great.

The tape's running out. So I'll stop now. One more month to go. One more month.

Amy

Sunday, May 11, 1986

Sarah's been taking a lot more of our time and attention lately. She wants to be read to constantly. She's always toddling over to Larry or me, holding a book in her hand. She has two favorites: *Goodnight Moon* and *Green Eggs and Ham*. She can listen to them over and over again for hours. But they do start to get a little boring after the fifteenth or sixteenth reading.

It's getting harder for me to keep up with Sarah because of how tired I've been feeling. My nausea's just about all gone but I'm always so tired, all I want to do when I have a free minute is go to sleep. And Sarah isn't very happy about that. She doesn't like to see Mommy in bed. Lord knows, she sees so little of me, at least I ought to be able to play with her when I do manage to get home from work.

Things have been very stressful and aggravating for me lately. First, it was the month in the NICU. God, that was terrible. I never want to spend another night in there! As far as I'm concerned, neonatology is a complete waste!

And when I finally finished getting aggravated in the NICU, it was time to start fighting with the chief residents and the rest of the administration about maternity leave. I know this might sound like a very old story, but it looks like they're trying to screw me again. Last week, they sent out these things called "Schedule Request Forms" for next year. We're supposed to fill them in and make requests for when we'd like our vacation time. Since I'm already in my third month of pregnancy and some people have already started asking me whether I'm pregnant, I figured it was time to bring the issue up with the chiefs. So last Tuesday I went up to their office and had a little talk with them.

Jon and Arlene were sitting at their desks and I walked in and said, "Hi. Guess what? I'm three months pregnant. I'm due next November and I need to arrange maternity leave." Just like that. You should have seen the looks on their faces. I thought Arlene was going to fall out of her chair. Of course, whether I'm pregnant or not doesn't really make any difference to them; they're not going to be here after June 1, so it isn't their problem, it's the new chiefs' problem. But they certainly were stunned just the same. It took them about five minutes to recover enough to congratulate me. In some ways it was worth suffering through these weeks of nausea and exhaustion just to see the looks on their faces!

Anyway, what's been done in the past is that a three-month period of maternity leave has been created by taking the one month of vacation and the one month of elective without night call we're entitled to [**residents have a one-month period called free elective; during this month, because they have no night call, they can travel to other**

cities to do electives] and adding another month of elective with night call. I told the chiefs that that's what I wanted to arrange, but they immediately started giving me a hard time. They said that as far as they knew, all I could get was my one month of vacation. They were unwilling to give me the other two months; they told me three months off would cause enormous problems in the schedule, and they just couldn't afford to do that. I've found out since then that three other women are pregnant and expecting at about the same time I am, and if they give us all two or three months off, it just might destroy the schedule.

Well, I don't care if my having a baby does destroy their precious schedule. I'm finished worrying about everybody else. What it comes down to is they're trying to discriminate against me and I'm not going just to sit back and stand for it this time! I'm tired of being pushed around and doing things just because the chief residents or the attendings or someone else tells me that that's the way it's got to be! So I made a call to the CIR [**Committee of Interns and Residents, the house officers' union**] office in Manhattan. I got the vice president on the phone and asked him exactly what the policy was for maternity leave. He told me that I'm entitled to six weeks of leave above and beyond any vacation or elective time I might have coming. That means that what I'm really entitled to is six weeks of maternity leave, four weeks of vacation, four weeks of elective without night call, and another four weeks of elective with night call. That's a total of four and a half months, and you can be sure I'm going to take all of it! If they hadn't tried to screw me over in the first place, I would have settled for three or maybe even two months. Now I'm going to take four and a half months, and if they don't want to give it to me, I'm just going to file a grievance with the union!

I went back to tell the chiefs about this on Thursday and they said they'd have to talk with the higher-ups before they could give me an official response. I got a call on Friday from Mike Miller's secretary, setting up a meeting for tomorrow. It looks like I finally got some action! After all this time, I finally figured out how to get things done around here. It's too bad everything has to be done through threats, though.

Wednesday, May 14, 1986

On Monday I met with Mike Miller about my maternity leave. He was very nice about the whole thing. First, he hugged me and told me how happy he was to hear the news. He seemed really sincere about it. Then he told me how upset he had been when he found out what the chiefs had tried to do to me. He said that in our program, we've always been very liberal about maternity leave. He went into a whole lecture about how we as pediatricians were supposed to be advocates not only for children but for their parents as well and how it would be hypocritical for us not to allow the residents proper time to be with their newborns. He was being nice to me, so I didn't start up with him about breast feeding and how it was hypocritical that none of us could ever breast-feed our children because we weren't provided with the proper facilities to ensure its success. Anyway, he told me he would guarantee that I got at least two months of leave after the baby was born and that I'd have easy rotations both the month before I was due and the month I came back after the baby was born.

Mike's always been nice to me, and I think he really meant what he said. I don't think he was saying those things just to try to prevent me from filing a grievance with the union. So I'm going to take his word for it. I'm going to trust him and I'm not going to speak to anyone from the union, at least not until next year's schedule comes out at the beginning of June. But you can be sure, if two months of leave are not written into that schedule for November and December, I am going to be on the phone to the CIR so fast it'll make your head spin!

I was a little distracted when I went in to talk to Mike because I've been really worried about my father. I called him on Sunday night and right near the end of this very nice conversation, he happened to mention to me that he had passed some bright red blood with a stool earlier in the day. Just like that, very matter-of-factly he said it. I immediately got upset and I asked him what he was going to do about it. He said, "Nothing."

I almost went crazy! He always does this to me. I yelled at him that he had to go to see a doctor the next day, and if he wasn't going to make an appointment, I would call his doctor and make the appointment and then come out to New Jersey and drive him to the doctor's office to make sure he got there. I think he got the message because he said he'd try to make the appointment but told me that his doctor was a very busy man and he might not be able to get to see him for weeks.

Anyway, right after I came out of Mike's office, I called my father at work. He said he had been able to make an appointment for this morning. He called a little while ago to tell me that the doctor had seen him, had done a full examination including a sigmoidoscopy [**an examination of the sigmoid portion of the large intestine**], and that the bleeding had been due to hemorrhoids. Everything else was fine. So I felt a lot better. My father told me I had made him worry for nothing. I told him that we didn't know it was nothing until we checked it out. It might have been something, and if he had ignored it, it might have cost him his life.

The doctor did tell my father that he did have to have the hemorrhoids removed as soon as possible because there was a possibility that there could be massive bleeding. He said he could arrange for the surgery to be done next Monday, and amazingly my father agreed; he's not admitting it, but he must either be in a lot of pain or really be scared about this. Whatever it is that's causing it, I'm glad he's acting so reasonably about it. But now I have to try to get next Monday off so I can go out to New Jersey to be with him. It shouldn't be too much trouble. Then again, you wouldn't expect it to be too much trouble to get a day off to stay with your baby who has the measles and a fever of 103, would you?

I've been feeling better over the past few days. My nauseousness is completely gone, and I'm not so tired anymore. I think I've made it through the worst part of this pregnancy. When I was pregnant with Sarah, once I made it through the initial yucky part, I felt wonderful until about three weeks before I delivered; then I felt like a blimp and couldn't move at all. So it should be smooth sailing for me over the next few months. There's only one more month of this internship left. I haven't enjoyed most of this year; maybe I'll be able to salvage this last part of it.

Tuesday, May 20, 1986

They did it to me again! I can't believe it! Yesterday was the day my father had his surgery. I had been trying since last Wednesday to get the day off. I thought it would be easy. I should have known better.

There were lots of problems from the beginning this time. First of all, I have clinic Monday afternoon, so I went to the director of the clinic and asked if I could switch my patients to another day just this once. I'd never asked her for a single favor in the past and I'd done lots of things for her like covering for other doctors who were sick. She hemmed and hawed for a couple of minutes and then when I explained that I needed to go to be with my father who was having an operation, she said she'd see what she could do. She called me on Thursday to say that she'd tried everything but there was no way she could cancel or reschedule the patients.

Fine. I understood what that meant. What that meant was that I wasn't going to be able to approach this whole thing as a grown-up. I was going to have to call in sick and lie about it. And because of that, I didn't even go to talk to the chiefs about switching my night call. I'd just call in sick and leave them hanging in the wind.

Well, yesterday came, and that's exactly what I did. I called everyone I had to, the chiefs, the clinic, and the emergency room, and I told them I had gastroenteritis and I couldn't make it in. Then I got dressed and got ready to leave for New Jersey. But just as I was about to leave the apartment, the phone rang. It was Mike Miller. He asked me if anything was wrong. I thought for a minute: Should I tell him the truth, or should I continue the story I made up? I decided to tell him the truth. It was obviously the wrong decision, because after I got finished telling him about my father, he said that they were strapped today, that a lot of people had called in sick, and if I didn't come to be on call at least that night, there would be only one person in the emergency room from five until midnight. So he made me a deal: He told me he'd let me take that day and the next day off if I came in to be on call that night. He said they were depending on me. He was so nice and so straight about it, I didn't see that I had much of a choice. So I went out to New Jersey and sat outside while my father had the surgery and stayed in the recovery room. I had to leave to come back to the Bronx just as he was going back to his room. It was so stupid; I didn't get to spend any time with him at all.

At least I knew he had done well during the operation. But all during the night, while I was seeing patients in the emergency room, I kept thinking how stupid this was. I mean, my father's not a young man; an operation like that can cause complications. And he had nobody else in the world to stay with him except me. So what was I doing while he was waking up from the anesthesia? I was seeing kids with runny noses and ear infections. I didn't need to be there; Evan Broadman, who was the senior resident, could have seen everyone by himself.

Well, I spent all day today at my father's bedside. He's in a lot of pain. The surgery's very uncomfortable. He didn't say anything about my not being there last night, but I knew he would have liked me to have been there. And I would have liked to have been there, too.

Monday, May 26, 1986

I was trying to think last night if I'd learned anything this year. I was taking care of a six-year-old girl who had been hit by a car. She had about a five-minute loss of consciousness but seemed to be fine by the time she reached us. I handled the entire case myself. I started an IV and sent off a CBC and a set of lytes [**examination of blood electrolytes**]. I got skull films, which were negative; I did a UA and found some blood in her urine, so I arranged for an IVP, which also was negative [**any patient who has had significant trauma and is found to have blood in the urine must have an intravenous** pyelogram, an X-ray evaluation of the kidneys, to make sure that no damage has been done to the kidneys]. I cleaned out her forehead laceration and sutured it myself. Then I called the intern up on the floor and admitted the patient for observation. I think that was pretty amazing! Considering I couldn't start an IV or put in a suture when I started last July, I think you'd have to say I've come a long way. It's funny, though; you never see it that way while you're doing it. While you're working, you're only aware of the things you don't know, not the things you do know.

Mark

Thursday, May 1, 1986

My mother told me I should always try to find something nice to say about a situation. I started my rotation at University Hospital last Monday, and ever since, I've been trying to figure out something nice to say about the place. I finally came up with something: The food is good. No, that's not even exactly true. It's not actually good, it's just plentiful. Plentiful and easily available and free; they give us meal tickets so we can eat three meals a day. And that's it. Outside of the food, I haven't found anything I've liked at University Hospital.

I've been on call one night and so far I've had one patient die. It was a patient I'd met before: the kid I saw in the West Bronx ER a few weeks ago who got bitten by the horse. Her name was Melissa Harrison, and she had this horrible disease, metachromatic leukodystrophy. She'd been going downhill for a while. She came in on Monday in status epilepticus **[the** state in which constant seizures are occurring]. Dr. Ruskin, her neurologist, came in and spent about an hour and a half talking to the parents. At the end of the meeting Ruskin came out and told me they'd decided that this was going to be it. We weren't going to do anything heroic, just fill the kid with enough morphine to keep her comfortable and then wait for the end to come. The end happened to come when I was on call Tuesday night.

This wasn't exactly the most comfortable situation I'd ever been in. I mean, I didn't know this kid from a hole in the wall. And here I was, being called on to stand by her bed and let her die without doing anything to prevent it from happening. Ruskin might have felt comfortable being in that situation, but she wasn't standing there at the kid's bedside. I was, and I felt pretty bad about the whole thing.

This kid's mother was a saint, though. I guess she saw I was pretty uncomfortable, and she spent a lot of time trying to calm me down. She told me about what Melissa had been like before she started going down the tubes. Isn't that wonderful? The mother of this dying girl had to spend the last minutes of her daughter's life calming down the intern who had gone completely out of his mind. Well, listen, it isn't completely my fault that I'm berserk; I'll be the first to admit that I might not have started out this internship with a full complement of marbles, but most of the berserkness I've been demonstrating recently is the result of the deep frying my brain's been receiving over the past few months.

Anyway, Melissa's mother was really great. She's a real Mother Teresa type. I can only imagine what kind of hell her life's been over the past few years.

So that was a great way to start out the month. I'm on

again tomorrow, and since I seem to have become the Intern of Death, I wonder which one of my panel of patients will be tomorrow's selection in the Meet Your Maker sweepstakes. Will it be Nelly, the three-year-old with AIDS who has PCP [**pneumocystis carinii pneumonia, a common cause of death in patients with AIDS**]? Will it be Jesus, the one-year-old with yet another bizarre metabolic disease, the name of which I can barely pronounce? Will it be one of the parade of renal transplant patients who are constantly marching onto the ward to get treated with medication that might stop them from rejecting their transplanted kidney? Or will it be a completely different patient, one I haven't even met yet, one who's waiting in the wings to make my life completely miserable over the next forty-eight hours? Only time will tell. And I don't think I want to know.

I'm going to sleep now. Maybe I'll sleep through tomorrow and the entire next two months, and when I wake up, I won't be an intern anymore. I can always hope!

Sunday, May 4, 1986

Great news! I was on Friday night and no one died. Nobody; no patients, no nurses, not even me! At least if somebody did die, I wasn't told about it.

Actually, Friday night was nice, if any night spent in any hospital can be called "nice." I didn't get a single admission. I even got five hours of sleep in University Hospital's very lovely intern on-call room. The on-call room is in reality a closet with furniture; it's about six feet by six feet and it's got a door, a telephone, and a cot. When they were building this hospital, they obviously decided to spare no expense when it came to the comfort of the interns. I shouldn't complain, though. I heard that as of four years ago, the interns didn't even have this closet to sleep in. They had to sleep in empty patient beds. That's always very dangerous, especially here at University Hospital, where there's an actual blood-drawing technician. There's always the chance the tech will find you lying in bed some morning, mistake you for a patient, and suck out all your blood.

I discovered another good thing about University Hospital. There's this porch attached to the cafeteria that you can actually go out on and get some sun. Actual sun in the Bronx! Anyway, I found this porch at lunchtime on Friday and I spent an hour out there on Friday afternoon. It was really beautiful. The weather's been great all weekend, too. The temperature's been in the seventies. Yesterday Carole and I went to this inn about an hour north of here. It was great, really relaxing, and we weren't caught in a rainstorm, a monsoon, a tornado, or any other natural disaster. Amazing! Maybe my luck is actually changing. Nah, it probably was just a fluke.

I don't really have too much to say tonight. I just wanted to show that it's still possible for me to be in a good mood. See—there's hope for me yet.

Tuesday, May 6, 1986, 9:00 P.M.

I was on last night. What a good time I had! What a wonderful learning experience it was! I had such a good night last Friday, I thought I was actually going to like the rest of my month. I thought it was going to be really quiet and restful. Then I was on last night and now I feel as if somebody dumped a fifty-pound bag of excrement on my head.

And I feel better now than I did a couple of hours ago! At six o'clock I was a genuine basket case! I was ready to manually extract the spleens of each of the chief residents without

the use of anesthesia. But then I went over to my grandmother's. She fed me a nice dinner and calmed me down.
Thank God for Grandma! Thanks to her, the chief residents
will live another day.

When I talk about it, I don't think it'll sound like last
night was all that bad. I mean, I had six admissions, which is
kind of bad, but all of them were electives and none of them
was sick, so it should have been pretty easy, right? It would
have been easy had they all come in at a reasonable hour. It
would have been easy had at least of few of them come in at a
reasonable hour. Did any of them come in at a reasonable
hour? Of course not! Why would anyone expect a kid who's
scheduled to have surgery the next day, who needs to be seen
by residents from at least three services [**pediatrics, surgery,
and anesthesiology**], and who needs to have blood work
and all kinds of other tests done, to come into the hospital
before nine o'clock at night? What a silly idea that is!

Well, anyway, they started to arrive at about seven-thirty
and they continued to show up until nearly midnight. I
couldn't believe it: A six-year-old who was scheduled to have
a T and A [**removal of tonsils and adenoids**] this morning
didn't show up until midnight. A normal six-year-old
shouldn't even be awake at midnight, to say nothing of a six-
year-old who's scheduled to have an operation a few hours
later! I was pissed, the anesthesiology resident who came to
see the kid was pissed, the surgery resident was pissed, everyone was pissed except the kid and his mother, who couldn't
understand what we were all so upset about. To them a six-
year-old coming in for an elective procedure at midnight was
completely natural.

So it took me until after two-thirty to finish all my scut
work on six lousy electives! And of course just when I was fin-

ished and I should have been able to get to sleep, Nelly, my AIDS kid with pneumocystis carinii pneumonia, decided to try to die on us. Boy, how happy I was to see that! It's me and Diane Rogers [the cross-covering senior resident] in a hospital that doesn't have a pediatric ICU, trying to keep alive a kid who's trying her best to get to heaven. It was amazing: She was perfectly fine one minute, and the next minute she was dropping her pulse to sixty and her blood pressure to sixty-five over forty-five. It really looked like the end was near. We stood around scratching our heads for a couple of minutes, trying to figure out what the hell was going on and what we should be doing about it. Her blood gas was still okay, so we knew it wasn't a ventilatory problem. Diane finally figured maybe we should try some Dopamine [a drug that increases blood pressure, among other things] to see what that did. I didn't understand the reasoning (of course, there isn't much reasoning I do understand), but the Dopamine seemed to do the trick. Nelly was good as new after that.

So there we were, with a kid with AIDS and PCP, who was going into shock, getting a Dopamine drip while on the regular ward. I had to stay with her for the rest of the night. I didn't get any sleep, and then I had to start rounds so I could have my usual morning fight with the blood-drawing tech who was refusing to draw blood on everyone. This has become a regular part of my day, I've kind of become addicted to it. Fighting with the blood-drawing tech is like drinking coffee.

So much fighting goes on at this hospital, it's unbelievable. Working at University Hospital is definitely like being drafted into the army during wartime. It's us against them, with the "them" being everybody who's not a house officer:

the attendings, the nurses, the lab techs, and especially the patients and their mothers. Work rounds in the morning are more like a pre-battle strategy session. We plan out the tactics we're going to use that day. But there are a lot of situations you can't plan for; things like sneak attacks. They tend to keep you on your toes.

Friday, May 9, 1986

I really don't know what to make of this place. This hospital definitely has some schizoid tendencies; sometimes it seems like the nicest place in the world. There are some afternoons when it's so peaceful and quiet, you can relax, sit out on the sun porch, even take a nap. And then there are some days where the patients all get sick at once, there are millions of admissions, and all you do is fight with everybody you can find. Take yesterday afternoon. At about three o'clock, the whole team got stat-paged to the adult ICU. We had one patient in there, an eight-year-old who had been hit by a car a couple of days before and had been unconscious ever since, so we all were sure she had arrested. We went running into the ICU and found she was fine, but the neurosurgery attending and one of his residents were standing by her bedside. As soon as we pulled up, the attending started yelling about how poorly we were managing the patient and how embarrassed he was that a patient who had been referred to him was getting such lousy care. She wasn't getting lousy care, she was getting great care. We all knew that. It's just that this guy has this quota: He has to yell at at least one house officer a day.

It was really hard to keep a straight face while this guy was yelling at us because he was sucking on a lollypop the whole time. It's hard to take this neurosurgeon seriously in the first place, but when he's got a lollypop in his mouth, it's damn near impossible!

And that wasn't even the end of it. Today, when we were on work rounds, we ran into the neurosurgery team. In spite of how poorly we had managed the kid, she had awakened out of her coma last night and seemed just fine today. Now her prognosis is excellent, and the neurosurgery chief resident told us we had done a great job with the patient. The attending immediately yelled at him, saying, "How can you tell them they did a great job less than twenty-four hours after I yelled at them for doing a lousy job?" The chief resident apologized and told him that since he hadn't been on rounds yesterday afternoon, he didn't know the attending had yelled at us. Then the attending got real pissed and said, "Next time I yell at somebody, I want the whole team there. I don't want to have to yell at people twice for the same thing!"

So last night I had a really quiet night. No admissions, just some coverage, and almost everyone remained stable. This was mainly due to the fact that Nelly, the AIDS kid, got transferred over to the ICU at Jonas Bronck. When Al Warburg, the daytime senior resident, found out that we had a patient on the ward on a Dopamine drip, he picked up the hot line to the chief residents' office and told them they had to transfer Nelly. So after I had been up all night with the kid, she got whisked away to Jonas Bronck. Don't get me wrong, I'm not complaining.

I guess the sickest kid on the ward right now is José, a one-year-old with this weird metabolic disease called argininosuccinicademia. The name of this thing is longer than the kid is! Anyway, having José on the ward is like taking care of an unremitting Hanson. He's constantly crumping and then stabilizing and then crumping again. He's lived in the hospital for the past couple of months, and all the nurses have come to love him. That's always a bad prognostic sign.

This disease has something to do with the urea cycle, and the kid is being treated with all these weird chemicals that make him smell really strange. I spent a few minutes standing at his doorway yesterday, sniffing his bouquet, trying to figure out what in hell it was he smelled like. It took a while, but it finally came to me: He smells just like the bottom of a birdcage. The kid smells like parakeet droppings! It's the strangest thing, but that's exactly what it is. Since I figured that out, I've become fixated on thinking of him as a parakeet. I'm waiting for him to start singing. And I'm sure it won't be long before he sprouts wings and just sort of flies away.

When you start sniffing the patients, I think it's safe to say you've been an intern too long. I think it's time to get out of here!

Sunday, May 18, 1986

Well, I haven't recorded anything for over a week, and nothing much has happened. Working at University really isn't so bad if you like taking care of kids with diseases whose names you can't pronounce. It's not like the other hospitals; they actually hire people here to do some of the scut work we're normally expected to do. So workwise there just isn't that much to do. But you more than make up for it in aggravation.

This is definitely the weirdest place I've ever worked in! At all the other hospitals, you really know what the score is. The rules are simple: They try to pile as much shit on your head as they can until you collapse, at which point a chief resident comes along, pats you on the shoulder, and gives you the weekend off to recover. Here the work isn't that hard, but you always have the feeling that you're missing something. You don't have control over anything. There are always attendings around who are trying to do things without telling you, and the parents always know more about what's happening with their kids than we do. It's very frustrating.

Nobody wants their kids touched by an intern. The parents all want the private attending to come in and draw the blood or start the IV. That's pretty funny because most of these private attendings haven't started an IV on a kid in years. People always naturally expect the more senior people to be able to do everything better than the interns. I'll tell you, at this point in the year there are very few people who are better than the interns at starting IVs, doing spinal taps, drawing blood, doing any kind of scut. But the parents still want to know why the private attending isn't coming in to do the stuff. So even though there's lots of time to sit out in the sun, I think I'd rather be in the wasteland of Jonas Bronck.

Well, there's only a little over a week to go and I'll be out of here. And then there's only one more month of internship left. That's pretty unbelievable, but I'm finding the idea of me being a resident even more unbelievable. In a little over a month there is going to be a group of poor, innocent interns who are actually going to look up to me with respect. They're even going to think they can trust me! My God, what a frightening thought!

Wednesday, May 28, 1986, 8:30 P.M.

Well, it looks like I made it. I just came home, which means I'm done with University Hospital. The rest of my internship consists of one measly month in the NICU at West Bronx. It'll be a cinch compared with last night.

I had the feeling last night that what was going on wasn't real. I figured this had to be a setup for *Candid Camera*. But nobody told me to smile, and no short, fat, bald guy came out and shook my hand. So I think it must have been real.

I was supposed to be on with Diane Rogers, but she called in sick and there was nobody to cover for her. So the chiefs asked if I would mind working on the ward by myself. Me mind covering a ward filled with twenty-five sick kids by myself? No, no way I'd mind it. I told them I looked forward to challenges just like this, that I welcomed just this type of adversity. In fact, I even told them I'd be happy to work every night next month by myself because that's the kind of guy I am. I don't think they realized I was being cynical, because somehow at around ten o'clock last evening I found myself rounding on the ward by myself. I even yelled at myself a couple of times for not following up on some scut I was supposed to do.

Anyway, everything was going fine, mostly because there hadn't been any admissions in a couple of days and the place was really quiet, but then at about four o'clock this morning I got a call from one of the neurology attendings, who told me he was sending in a kid with a brain tumor who had been in status epilepticus for about four hours. Status for four hours! I told him fine, I welcomed these kinds of patients, that I looked forward to challenges just like this, and that I'd be waiting for him. Then I calmly hung up the phone, ran for the staircase, and started moving in a downward direction. I was getting out of there; I might be crazy, but I'm no fool.

When I hit the third-floor landing, something weird happened. I got this sudden rush of guilt and I realized I'd have to go back. So I slowly climbed back up, told the nurse what was happening, and got ready.

The kid got there at about four-thirty. He was seizing, all right, there wasn't any doubt about it. I had no trouble figuring out he was seizing; what I had trouble figuring out was what I was going to do about it. So I called the neuro attending at home, and the first thing he did was yell at me for waking him up. I was expecting that; the first thing everybody does when you call them from University Hospital is yell at you. But then I asked him what I should do, and he said, "You've got a kid who's seizing. What the hell do you think you should do?"

My neurons turned on and I waited a couple of seconds for an answer to come out of my mouth. When it finally did, it was, "Give him an anticonvulsant?" The neurologist said, "Brilliant," so I knew I was on the right track. I said, "Should I start a line, give him some Valium, and then load him with Dilantin?" He told me that that sounded like a wonderful idea, so that's what I did. I got the line in, I pushed the Valium, and the kid suddenly stopped seizing. It was great. By six-thirty I had him stabilized, lying in bed, sleeping, which was a lot more than I can say about myself.

This was pretty amazing. I have a lot of trouble believing I was capable of working by myself for a whole night and even admitting a seriously ill patient and not making any major screw-ups. I guess now that I've got pediatrics perfected, it's time to try another field. Maybe I'll become a heavyweight boxer.

Bob

At the beginning of the month, Amy Horowitz told the chief residents that she was pregnant and needed to arrange maternity leave. Her announcement was met with the release of an explosion of venom aimed at her by the chief residents, who weren't about to give any special treatment to Amy just because she happened to be pregnant. This reaction of the chiefs produced the release of an equal explosion of venom from Amy, who, fed up with what she viewed as the chronically poor treatment she'd received all year long, decided to call the Committee of Interns and Residents to find out exactly what she was entitled to. The situation, which was escalating, was finally defused by Mike Miller, who managed to put Amy's ire to rest, at least for the time being.

The issue of maternity leave for house officers is a relatively new one. In the 1950s, residency training programs didn't have to worry about developing specific policies regarding leaves of absence for new mothers for two reasons: First, at that time, there were very few women in medicine;

and second, many programs strictly prohibited house officers of either sex from being married. Over the past thirty-five years, however, this situation has changed dramatically: Today over 50 percent of the 105 house officers who make up our program are women, and the majority of these women are married. In recent years we've averaged about five new babies born to female house officers annually. As a result, a definite plan regarding maternity leave has been developed, with the intern or resident receiving about three months away from the hospital around the time of delivery.

The development of this plan has been met with mixed reactions from the house officers, both male and female. After all, if one person is given three months off, someone else is going to have to fill in for her. An attempt is always made to spread the coverage evenly, but often a few people wind up doing what they consider more than their fair share. This leads to resentment directed toward the person on maternity leave, resentment that may stay with her through the rest of her training.

But the problems that female doctors face are certainly not limited to these issues surrounding maternity leave. Discrimination against all women in medicine is rampant. Although the foundation of this discrimination is rooted in the past, when medicine was exclusively a male profession and when house officers were referred to as "the boys in white" and specialists such as ear, nose, and throat surgeons were called "ENT men," the image lives on in the public's mind. It lives on mainly because the medical establishment, which at this time is composed of those "boys in white" of the 1940s and 1950s who have grown up and taken charge, perpetuates the myth. And so the acceptance of women as medical equals of men is a difficult goal to attain.

It's easy to see examples of discrimination. In our emergency rooms, any male who has contact with a patient is immediately referred to as "Doctor" by the patient's parents, regardless of whether he is a doctor, a nurse, a medical student, or a clerk. Any female, no matter how senior or expert, is automatically assumed to be a nurse. At the beginning of the year, the female interns take great effort to correct the parents; they explain that they've gone to medical school, have graduated, and are just as much doctors as any man; but as time passes and it becomes clear that these explanations are doing little to change the public's conception and actually are creating hostility between doctor and patient, the women try to ignore what they consider this slight, managing just to cringe a little and swallow hard a few times when it happens.

And patients often believe that women can't do as good a job as men when it comes to the technical aspects of medicine. I've seen it a hundred times: parents refusing to let the senior resident, who happens to be a woman, draw blood, do a spinal tap, or start an IV on their child, demanding that the male doctor in the next examining room, who happens to be an intern, try the procedure first.

But the patients clearly are not the only ones who discriminate against female doctors; it's also firmly entrenched in academic bureaucracy. Thus far, few women have achieved positions of authority at medical schools in the United States. As an example, only a handful of the chairmanships of pediatric departments, the specialty with the largest percentage of practitioners who are female, are held by women. Part of this is due to the fact that until recently there weren't many senior physicians who were female, but part is definitely because qualified women are frequently not offered a job when an equally qualified male candidate is available.

Also, it becomes difficult for female doctors to deal with nurses, the majority of whom also are women. A good intern has to be aggressive, but aggressiveness is not a trait that is viewed as acceptable in women. When a male doctor orders a nurse to perform a task for his patient, it is viewed positively; he is just carrying out his responsibility. When a woman is the one who requests that a nurse do something, she is regarded as "uppity" and a troublemaker. It's a bind that is difficult for the female house officer to resolve satisfactorily.

These issues present an enormous identity problem for the female intern. On the one hand, she's not getting equal treatment from her patients or from the nurses; on the other hand, she has few or sometimes no role models to guide her in her training. Very often this second problem is more serious than the first.

Take Amy's problem as an example. Amy has done an amazingly good job. She has worked for an entire year as an intern, fulfilling all or most of her responsibilities. But at the same time, she's also had to be a mother to Sarah, trying her hardest to fulfill the responsibilities of what clearly is a second important full-time job. She's done all of this without anyone pointing the way for her; there are few faculty members around who could share their experiences as an intern and a mother with her. And although she's had some help, mainly from her husband and her baby-sitter, she's found little support within the system. The chief residents never wanted to know how sick her daughter was or what family obligations she had; they weren't even happy or excited when Amy told them that she was pregnant; they only wanted to know that she'd be at the assigned place at the assigned time and that her job would get done.

And there are very few options open to female residents

with children. The attitude is basically this: If you want to have a baby and you want to spend time with your baby, you should take a year or two off; if you want to work, you should put off childbearing until after residency training is completed. A happy medium—that is, working as a house officer halftime and spending the rest of the time as a mother—is at present available at very few hospitals.

Changes are occurring, but they're occurring slowly. Eventually the young women who are house officers today will move into positions of authority, and the concept of medicine as a private club for men will gradually fade away and ultimately die out. At that time, a more realistic attitude toward women in medicine will evolve. And innovations such as shared residencies with two or more people fulfilling the responsibilities of one house officer, day-care facilities within hospitals, suitable facilities to encourage breast feeding, and fair maternity leave policies, which today are considered radical and expensive luxuries, will become commonplace. But as of now, Amy and her sisters in medicine must bear a heavy load.

Andy

Tuesday, June 24, 1986, 5:15 P.M.

My internship ends in three days. I'm moving back to Boston on the twenty-eighth. I can't believe this is finally going to be over so soon.

This has been a tremendously long year, in some ways feeling more like three separate years than just one. The first year stretched from when we started back in June to when I finished on Adolescents' at the end of September; that first period took me from the time when I was enthusiastic and up about medicine to the point where I reached my first real depression. The second year included University Hospital and my first three months at Jonas Bronck; this was the best time for me. I was "up" for a lot of it, I managed to get myself organized, I pulled some things together for the first time, and I really began to see that the experience was eventually going to turn me into a doctor; the time I spent on the east campus was the most optimistic period for me.

The last period, which has been the most difficult, took up about the past four or five months, from the time I first walked into the PICU until now. I've gone through hell these past five months; I became emotionally wrecked, much worse than I ever thought I could. It's affected every aspect of my life, including my relationship with Karen, which I've always thought was unshakable. There was a time earlier this month when things had gotten so bad that we were seriously considering splitting up. This last period of internship has turned me into a very selfish and self-centered person. Thank God I've gotten some insight into what's been happening. I think Karen and I have patched things up pretty well now, but it was very disturbing there for a while.

The hardest period of this year happened during the last half of May. I hit the big burnout. I really didn't give a shit about anything; all I wanted was to be left alone by everybody. This lasted through the first couple of weeks of this month. At one point about two weeks ago, our attending sat me down and said, "You know, Andy, when you go to that new institution, it's going to be very important for you to make a good impression during the first couple of weeks. Everyone is going to judge you for your entire stay there on how well you do at the very beginning. So snap out of this!" He realized I was just going through the motions, and it was nice of him to talk to me about it. I've pretty much recovered from that burnout now. I don't know how, maybe it was because of what the attending said to me or maybe I just kind of woke up and realized what was happening on my own, but now I can behave myself most of the time without cursing and being moody and driving everyone crazy.

Over the past week or so I've started listening to some of the tapes I made back at the beginning of the year, and I noticed something: It seems like I remember the bad things much more vividly than I remember the good. I've forgotten a lot of the good things, the successes, the patients who have walked out of the hospital and have said, "Thank you" and have shaken my hand. Those people have been crowded out of my memory by all the ones who died or who did poorly, the ones who wound up breaking my heart.

Internship is supposed to be an important educational experience, but I'm still not sure what I've learned. One thing I've accomplished this year is I've managed to develop my own personal style as a doctor. I've turned out to be more compulsive than I thought I would be. I've gotten very efficient; I'm more able to decide what's important and what's not than I was a year ago, when I don't think I really knew how to prioritize at all. And probably a year from now, I'll look back and realize how little I know about what's important right now. I also think I somehow managed to retain my sense of humanity and my sanity among the inhumane environment of the hospital and the insanity of everything we do and the craziness of the Bronx. Thinking about it like this, I guess I really did pick up a lot this year.

But I definitely don't feel ready to be a second-year resident yet. I don't feel ready for that next step, that sudden acquisition of great responsibility where I'm the one who has to make the decisions and oversee the interns. I've gotten pretty good at doing what I've been called on to do as an intern. I have my own opinion now about how things should be done, but I don't argue much if I disagree with the residents or the attendings. They've got their jobs to do and I've got mine.

The other day we got new medical students. Brand-new, green, third-years, who've never been on a ward before. Our resident took great pains to explain carefully everything that was happening to these guys, like what a FIB is and what tests were done in a CBC. I was bored to tears.

We were all on our best behavior during rounds, but as rounds were ending, the other interns all tried to impress the students with how jaded and how cynical they had become. I stood there for a while as this discussion began and I just thought, Listen to all this bullshit! After a few minutes I couldn't take it anymore; I didn't want to be a part of this scene. So I just walked off. This kind of thing, trying to impress these poor third-year students, gets old really fast.

But I had fun with my stud [**student**] the rest of the day. I caught him in the library reading at about noon and I said, "Give me a break! What are you going to do, put on a nice clean shirt and tie every morning and spend the entire day sitting in the library reading textbooks? You're not going to get anything out of sitting in the library." So I forced him to get up and follow me around. I showed him some of the ropes. This afternoon I asked him to write a progress note on one of the patients he picked up and he wrote one of the worst notes I've seen in my entire life. It's so funny. He had absolutely no idea what was expected of him or what was supposed to be written in the chart. It kills me because he seems to be so bright and eager to work, but he just doesn't understand how to do anything yet. So tomorrow I'm going to have to really start to teach him things from scratch. But it's so hard to try to get my mind back to where a beginning third-year student is. I just can't put myself in his place.

Friday, June 27, 1986, 8:00 P.M.

My friend Ellen always used to talk about the need to process what was happening to all of us. She told me recently that it wasn't until the last few months of internship that she's been able at least to start to fit some of the pieces together and begin to understand what had happened inside her. I guess I've been able to do that only a tiny bit so far. I'm still standing too close to things to have any real insight. There's a lot of my internship I haven't talked about on these tapes. There have been things that were just too painful to go into; they would have been too damaging to bring up at the time, and now I've forgotten a lot of the details. But they've had their effect on me.

I'd like to think that overall this has been a good year, but I can't. It has been good in the professional sense. I was transformed from a medical student into a doctor. I've learned a great deal about patient management and how to think on my feet while half asleep. I think internship did all that extremely well. Thank you, Schweitzer Peds Department. All of you helped me make that transition.

Internship was also good in providing the battlelike atmosphere that brought me close to a bunch of strangers, my fellow interns, and very close to a few people to whom I'll forever have a bond, no matter how infrequently we communicate, no matter how physically far apart we drift. In all other respects, though, my internship was a draining, dehumanizing, destructive experience. It's almost like we started out in July smelling of cologne and perfume, and dressed in freshly laundered formal evening clothes, well-mannered and even-tempered with warmth in our hearts and great expectations, but by the end of the year we had become tattered, unshaven, smelly, cynical, snarling survivors of a long and somewhat meaningless struggle with ourselves and the rest of the world.

Amy

Thursday, June 5, 1986

So far, this has been the best day of my internship. Today's the day of the Pediatric Department picnic; the attendings all cover the wards so we can all go out to some park somewhere and have a good time. That's not exactly what I decided to do. When our attending showed up and told us we could leave, I came right home, picked up Sarah, and took her to the Bronx Zoo. Just the two of us; it was the first time all year I got to be alone with her during a workday. I'm so glad I decided to spend the day with her instead of going to the picnic. I'm really missing the best parts of her childhood.

The Infants' ward is pretty much what I expected. In some ways it's like being in the NICU except there aren't any really tiny preemies around. There are a lot of babies who graduated from the NICU. The ones with any real chance of a normal life go home; the disasters come to Infants'.

Of three babies on the ward who are DNR's, I'm taking care of two of them. One is Kara Smith, an eight-month-old who got meningitis about four months ago. She spent most of February in the ICU upstairs; she had everything wrong with her, there were problems with every single organ system, and all the doctors who had anything to do with her were sure she was going to die. But she didn't die, and eventually they transferred her down to Infants', to the DNR room, where she's been living ever since.

It's really sad; she's completely vegetative; she can't do anything. She has no head control, she can't smile, she can't suck. The nurses feed her through a G-tube [**gastrostomy tube: a tube inserted through the abdominal wall and into the stomach; G-tubes facilitate feeding of children who are neurologically impaired enough not to be able to suck or swallow**]. Five times a day they squirt blenderized baby food into her, and an hour or so later they change her diaper. She also has a trach so she can be suctioned [**babies with no gag reflex will not swallow the normal secretions that build up in the back of their throats; as a result, if these are not removed mechanically, the children will choke**]. And pretty much, that's the extent of her care. Since she's a total DNR, we don't draw bloods on her for anything, we don't culture her if she gets a fever, and we're not supposed to start her on any antibiotics. Eventually she'll probably develop pneumonia and die. But it's already been four months and she hasn't gotten pneumonia yet.

One of the nurses who's really attached to her told me that Kara's mother used to come every day when she was first moved down here. Eventually she only came every other day, then a couple of times a week. Now she comes maybe once a week. I haven't met her yet; usually she shows up late at night, so I suspect some night when I'm on call I'll run into her.

My other DNR baby is Lenny Oquendo. He's six months old. He's never been out of the hospital, and it looks like he never will be. He was one of the NICU disasters; he weighed a little less than six hundred grams [**one pound, five ounces**] at birth and spent three months on a ventilator. He has a grade IV IVH, severe hydrocephalus, and about a dozen other problems. He also has a G-tube and a trach. Lenny's mother hasn't come to see him in months. She seems to have completely lost interest in him.

There's a third DNR baby in the same room, but he's Ellen O'Hara's patient, and I don't know much about him. But that room is so depressing! The nurses and the rest of the staff buy these kids clothes and toys and things to try to liven up the atmosphere. But it doesn't help, it only makes everything that much sadder; the clothes and toys only make you realize how different these kids are from normal children. Just going in there and seeing those three hopeless and helpless babies lying in their cribs, it makes you want to cry! But at least they aren't much trouble. The only thing we have to do for them is rewrite their orders once a week and remember to sign them out to the intern on call.

The rest of the ward is filled with assorted disasters. There are three babies with spina bifida who have shunt infections [**infection of the ventriculoperitoneal shunt, the device that drains fluid from the brain into the abdominal cavity**] and are getting IV antibiotics, there are two babies with infantile spasms [**a severe form of seizures**] who are being treated with ACTH [**the medication used in this type of seizure disorder**], there's a nine-month-old with AIDS who was in the ICU last week with PCP but who's getting better. There are even a few normal children who have bronchiolitis.

Working on this ward really takes a lot out of you. It's emotionally very taxing. So having today to spend with Sarah was especially good. It raised both our spirits.

I'm on call tomorrow night. I'm going to stop now and actually cook dinner.

Sunday, June 8, 1986

I've been in a good mood this weekend. The schedule for the next year finally came out on Friday. They actually came through with what they promised: I'm scheduled to have my CERC rotation [**a month spent learning developmental pediatrics at the Children's Evaluation and Rehabilitation Center on the east campus; CERC is a calm, non-stressful experience**] in October, my vacation in November, a month of elective without night call in December, and my neuro selective [**a rotation learning child neurology; like CERC, neuro is pretty laid-back**] in January. They gave me what I wanted. Finally, after everything that's happened this year, I wound up getting something without getting screwed!

It's hard for me to believe that I have only three more weeks of internship left. At this point in time, I'm fairly sure I'm going to be able to make it the rest of the way. I hadn't been able to say that before this week. I'd been dreading working on Infants' for months; I'd heard only bad things about it. But actually, although I can't say I'm really enjoying the patients I'm following, I am having a good experience here. We have a very good attending, Alan Morris. He's an excellent teacher and I've been learning a lot from him on

attending rounds. And we have a strong team: Ellen O'Hara
and Ron Furman are the other interns, and they're a lot of
fun to work with. And our senior resident is my very favorite
person in this whole program, Ben King. Ben's a little burned
out at this point; this is his last month of residency, and I
don't think he really wants to be in the hospital. Yesterday, on
work rounds, he got into a wheelchair and made Ron push
him around the ward. He's funny and he makes working easy
because he's got excellent judgment. So, probably for the
first time all year, I'm actually part of a team I like being on.

I was on call with Ben yesterday. It was a very quiet day. I
had only two admissions, an eight-month-old sickler with
dactylitis [**inflammation of the hands and feet, usually the
first painful manifestation that occurs in children with
sickle-cell disease**] who didn't require any work, and a nine-
month-old with bronchiolitis who was admitted from the
West Bronx emergency room but who Ben immediately sent
home. It was really funny: I went down to the ER to get the
baby at about three in the afternoon and he really didn't look
that sick. But I didn't question it, I just brought him up to
the ward. Then Ben came by to see him and he said, "Why
did they admit this kid?" I told him I didn't know. He listened to the baby's chest and said, "This kid doesn't have
bronchiolitis. He's healthier than I am! Send him home
before something bad happens to him!" Just like that. His
mother got him dressed and they left. I don't know any other
resident who would have done that. But if you ask me, it was
the right thing to do.

So all in all, I haven't been too overly stressed on Infants.
Calls haven't been bad, and I've been getting out at a reasonable hour: not three or four in the afternoon, but usually no
later than five. It's staying light out until seven o'clock now,

so when I get home I can take Sarah out onto the lawn in front of the apartment building and just sit out there with her. It's nice. It's too bad the rest of the year hasn't been like this.

Sunday, June 15, 1986

Kara Smith died Friday night. She had developed a fever on Thursday; Ron was on, he examined her, and he thought she had pneumonia. He didn't do anything about it, just wrote a note documenting it in the chart. Then on Friday during the day her breathing became very labored. She must have been hypoxic. I felt very uncomfortable. I kept coming into the room to check on her. I knew she was DNR, but just sitting around doing nothing really bothered me. I wanted at least to get a blood gas and maybe start some oxygen, but the rules are no treatment.

Then finally on Friday night, one of the nurses called me around midnight to come to see her. She was blue and gasping for air. Her heart rate was down to about forty, so I figured the end was near. I called the resident on call to tell her what was happening, and she came down and checked Kara; she agreed with me that she was dying. We didn't do anything; we just sat by and watched.

She finally stopped breathing at about twelve-thirty. We covered her with her blanket and just walked out of the room.

I called the mother. I had met her last Saturday night. I had been on call and she came in at about eight o'clock. She didn't say much to me, only that she was pleased to meet me. It wasn't much, but at least I knew who she was and she knew who I was.

I got her on the phone and told her that Kara had died. She didn't cry at first. She seemed very composed. She asked if I thought Kara had felt any pain; I told her I didn't think so, that she had seemed comfortable the whole time. She asked if I knew what the cause of death was, and I told her about the fever and the breathing problems and the fact that she had probably developed pneumonia. Then the mother said, "I guess she's up with the angels now," and that's when she started to cry. I couldn't think of anything to say; I just sat at the nurses' station with the receiver up to my ear.

When she finally stopped crying, she apologized to me. She told me she'd been prepared for Kara's death for months and that she didn't think she would cry when the time finally came, but that she just couldn't help it. She said, "They told me she was going to die and I came to accept it, but I never really believed it." She started crying again at that point, but only for a minute or so. After we hung up, the rest of the evening was quiet; I didn't get any admissions, and the ward was calm. I went to the on-call room but I couldn't get to sleep. I kept thinking about Kara's mother.

Tuesday, June 24, 1986

Tomorrow's my last night on call as an intern. I've made it! It's hard to believe, but I actually survived. Believe me, it's not something I'd want to do again.

Looking back at the year, there have been a lot of things I've disliked; I didn't like the way I was treated by the chief residents, I didn't like the fact that I had to be on call every third night, I didn't like being tired and exhausted all the time, and I didn't like having to take care of sick, sick children. But definitely, the thing I disliked the most was being away from Sarah. I know I've said it before, but it's still true: I've missed some of the most important moments of my daughter's childhood.

I've asked myself a lot lately whether I'd have done this

internship if I knew then what I know now. I'm not sure what the answer to that question is; over the past few weeks I've tended more toward, "Yes, I would do it." But there are some days, when things are very stressful, when the answer is, "No way!" I guess it's silly to ask the question, though. I mean, it doesn't really matter. I've done my internship, I'm finished with it, and I never have to do it again. That's all that's really important.

Mark

Sunday, June 1, 1986, Noon

I just got back from my first night on call in the neonatal intensive-care unit at West Bronx, and I'm really starting to get the feeling that I'm not going to become a neonatologist when I grow up.

I got to the unit a little before eight yesterday morning. I got sign-out from Elizabeth, who was on the night before. I wouldn't say she was exactly sad to be leaving. Then she left and I started running around, and I continued through the night. I ran to the labs, I ran to the DR, I ran to the babies who were trying to die. The only time I sat down during the entire twenty-four-hour stretch was when I had to write those endless, pointless progress notes that go on for pages. It's a total waste, me writing notes. It's definitely gotten to the point where I can't even read my own handwriting anymore. Anyway, the whole day was horrible. Yesterday made my month in the nursery at Jonas Bronck seem almost pleasant!

Monday, June 2, 1986, 8:30 P.M.

I just got home. It's eight-thirty and this is supposed to be my good night, and I just walked through the door. Oh, this is a nightmare. But do I care? No, I don't care at all. Why don't I care? Because I just stopped at the supermarket on the way home and found blueberries. When blueberries appear, the end of the year is nearly here. They can do whatever the hell they want to do to me, but I don't give a damn anymore. Because I've made it. I've made it to the blueberry season.

So what's life like in the NICU? It's wonderful, great, like a vacation in Cancún during the rainy season. The unit is really very small. You can walk from one side of it to the other in about ten or fifteen steps. But packed into those ten or fifteen steps are some of the sickest patients you could possibly imagine. It's one disaster after another. There's a roomful of preemies who don't do anything all day but seize and try to die; there's a roomful of cardiacs **[babies with congenital heart disease who are being evaluated for or are recovering from cardiac surgery]** who only rarely seize but who always are trying to die; there's a room of miscellaneous disasters; and a fourth room, filled with social holds.

And the unit continues to stay full, mainly because of the topnotch obstetric service. OB is run by a team of killer midwives who are really heavily into what they call "the psychosocial aspects of childbirth." What that means is, they encourage the mothers to hold their babies right after birth to make sure they bond, no matter what's happening to the kid. On Saturday, Eric Keyes **[a senior resident who was cross-covering the unit]** and I were called to the DR stat for fetal distress. We got there and found this tiny midwife pushing on the belly of an enormous pregnant woman. The midwife told us she was applying external abdominal pressure. This pregnant woman must have weighed at least three hundred pounds, and the midwife weighed ninety at most. As she pushed down, it looked as if the midwife was going to be swallowed up by the pregnant woman. Anyway, we looked at the fetal tracing and saw there were late decels [**late decelerations: a heart pattern indicative of fetal distress**], so Eric suggested maybe they should think about doing a C-section. The midwife gave him a look I was sure would instantly turn Eric to solid rock but apparently it didn't, because seconds later, when the membranes ruptured and meconium started splattering all over the room, Eric immediately said, "Holy shit, let's get ready to intubate!" A little while after that, this tiny baby came flying out. The midwife caught it, wrapped it in a towel, and immediately handed it over to the mother.

I thought Eric was going to blow out his cerebral artery right then. He looked at the midwife for a second, then he looked at the baby, who was blue and not breathing, and he yelled, "What the hell is going on here?" The midwife turned to him and said, "Bonding. Shut up and go away!" Eric immediately grabbed the baby away from the mother, brought it over to the warming table, and we started working on it. The kid wasn't breathing. Eric intubated and sucked out a huge glob of meconium and then we started to bag the kid [**blow oxygen through an endotracheal tube directly into the baby's lungs**]. The baby picked up at that point and cried for the first time since birth. His heart rate came up, he started breathing on his own, and he turned pink, which looked much nicer on him than his original blue. It looked like a save.

But that wasn't the end. Just as we were finishing, the midwife came over and started yelling at Eric, telling him his grabbing the baby away like that severely disturbed the mother-child relationship. Eric said something like, "Oh yeah, sure, anoxic brain damage would have markedly improved the mother-child relationship, right? Bonding to a blue baby is much better than bonding to a pink baby. How stupid of me to interfere." They then got into a real big shouting match, right there in the DR. Eric told me later the midwives are always like that. He said you can expect to get into at least one argument with them a night. He said he thought there must be a required course in blue-baby bonding in midwife school.

Thursday, June 5, 1986

Today was the Pediatric Department picnic. The people in charge actually gave us the whole day off just to go and have a good time. It's so out of character, it's almost frightening!

This morning at about eight-thirty, our attending, Laura Kenyon, showed up and told us just to sign out to her and get the hell out of the hospital. Elizabeth and I were out of there instantaneously! If an attending's offering, we aren't about to give her a chance to reconsider. So we drove up to this camp in Chappaqua where the party was going to be. When we got there, we couldn't believe it. It was acres and acres of green grass and trees. It was great!

The picnic was actually a lot of fun. Just about all of us were able to go. We played softball, ate hamburgers and hot dogs, and drank much too much beer; in other words, we did all the things normal people might do if they were on a company picnic. We did a good job of pretending we were normal, at least for a few hours. It gave us hope that someday we might be able to shed this schizophrenic outer coat we've grown and return to the Land of Normalcy.

Anyway, on the way to the camp, we passed Peter Anderson's house, the place where we had orientation almost one year ago. Boy, that's amazing! It's hard to believe that orientation happened a year ago. It seems more like something out of a different century. There were the Middle Ages, the Renaissance, and orientation at Peter Anderson's house, not necessarily in that order. Well, what difference does it make? In another couple of weeks a whole new group of interns will be deposited on Peter Anderson's doorstep, sweating bullets. I bet they're all sweating bullets right now. I remember last year at this time, I was scared to death. By the way, if you haven't guessed by now, sweating bullets was the completely correct reaction. I wouldn't trade places with those guys for all the money in the world!

I left the picnic at about three-thirty because I had to get Elizabeth back to the hospital. She's on call tonight. Elizabeth wasn't exactly in the best mood today. It's kind of hard to enjoy yourself when you know that in a couple of hours you're going to be face to face with your worst nightmare.

I'm going to watch some TV now. Yes, it's been the kind of day normal people have, and I'm going to end it the way normal people end their day. I'm going to watch *The Tonight Show*!

Twenty-three days to go. But who's counting?

Sunday, June 8, 1986

Carole and I are getting along really well. It's kind of frightening. Either I'm over my internship depression, or she's slipped into a serious state of depravity. Anyway, it looks like our relationship has weathered the year. I'm glad it did, I guess. I like Carole a lot.

I just got off the phone with Elizabeth. She's on call

tonight. She said her foot is feeling better, but it's still not great. I don't know if I mentioned what happened to Elizabeth last week. She was on on Friday night and there was a code about 3:00 A.M. She told me she was in another part of the unit, trying to teach one of the cardiac kids how to breathe like a human, when the alarm went off in the preemie room. She went running in there but tripped on an electrical wire on the way and flew about ten feet into the air. This is a new Olympic event, the Preemie Resuscitation Slalom Course. She must have made a perfect landing, because she said the judges gave her scores of 9.5 and above, but she came down on her ankle, which got all twisted up. A couple of hours later, after she had made sure the preemie who had coded would live to face another sunrise, she was drawing the morning blood and noticed her ankle was hurting. I got there about that point and we rolled down her sock and both noted that her ankle had become the size and color of a ripe eggplant. At about that moment she said she was feeling a little queasy. I noted that her face had turned a sickly shade of green. That was right before she passed out.

"Yes, I'll tell you, they just don't make these interns like they used to! At the first sign of adversity, they all find it necessary to fall over. They're just not as durable as they used to be in the Days of the Giants!"

Anyway, we got her a wheelchair and I took her down to X ray. There weren't any fractures. Laura Kenyon got hold of an orthopedic surgeon who examined her, said it was just a flesh wound, and wrapped her ankle in an enormous Jones dressing [**a bulky dressing made of three layers of Ace bandages**]. She was up and caring for the clients in less than two hours. What a trooper!

I'm finding it very difficult to concentrate on my patients.

They've all become a blur to me at this point. I get one preemie mixed up with another; all the cardiacs seem the same; I just can't keep them straight anymore. I think I've got spring fever. I'm going to stop now.

Saturday, June 14, 1986

It hasn't been such a bad week. There are these bugs [bacteria] flying around the NICU that seem to be resistant to every antibiotic known to man. I don't know how they got into the unit, but I'm glad they did, because it means we're contaminated and closed to all admissions.

If I had known closing the unit would have been that easy, I would have brought the bugs in myself. I must have some type of bacteria resistant to every antibiotic known to man living in my apartment. I seem to have everything else living here. I can see a great future in the bacteria-resistant-to-every-antibiotic-known-to-man mail-order business. Interns all over the world would beat down my doors trying to get enough bacteria to close down their particular ICU. What a great concept!

Anyway, the infection hasn't done my old patients any harm. Of course, these kids are so sick, it's kind of hard to tell whether something does them harm or not. But it has caused us to have a nice, leisurely week.

Saturday, June 28, 1986

Well, it's over. It's all over. I am no longer an intern. As of nine o'clock this morning, I officially became a junior resident. No more internship! Ever! No more daily progress notes! No more blood-drawing! No more IVs! No more fighting with lab technicians! No more fighting with elevator operators! No more mock-turkey sandwiches! No more patients who are as sick as Hanson! No more Hanson!!

I think you can see here that I'm exaggerating a little. I think you can also see that I'm completely out of control! And I don't care! Because I'll never have to be an intern anymore, never again. Hooray!

This morning at about eight o'clock, I was drawing the morning blood and whistling. Yes, I've been whistling on blood-drawing rounds over the past few weeks because it's such fun! Anyway, I'm walking around the unit whistling and jabbing great big needles into my wonderful patients because I love them all so much and this guy who looked lost and scared to death came in and asked, "Is this the nursery?" Guess who he was? He was . . . an intern. He was the intern who was on call in the NICU today! And I didn't know who the fuck he was! Because he's brand-new!

I told him he was in the right place and I showed him where to get a set of scrubs and then I showed him the patients. While I was doing this, I stayed between him and the door at all times because I was sure that at some point or other he was going to bolt, leave the hospital, and never come back, and I'd have to stay and be on call again. But he didn't leave. He was really nervous, but he seemed very enthusiastic. It was like I was talking to a member of a completely different species on the evolutionary tree. He took notes on this clean pad on this brand-new clipboard. He didn't ask any questions, and I'm convinced he didn't understand a single word I said to him.

Anyway, I finished rounding with him at about ten and then we all gathered in the West Bronx library and the party started. A bunch of us were sitting in there, drinking champagne and getting soused. At ten in the morning. We stayed until about eleven, when the bar across the street opened, and then we all went over there for brunch. It might seem strange that ten or eleven interns would be sitting around a bar drinking at eleven o'clock in the morning, but hell, we weren't alone. The place was packed! It wasn't only pediatrics that changed over today; medicine and surgery changed also, and everyone was getting loaded. Anyway, we stayed there until about two. I just came home to take a nap and get ready for the real partying, which will start tonight.

I thought when it was all over, I'd have all these great, profound thoughts about internship. I've been trying to think of something profound to say all day, but I can't come up with a single thing. Internship sucks, that's all there is to it. It just flat-out sucks. But hey, it's not my problem anymore. I'm no longer part of that lower class of humanity! I'm pretty sure that if you come to me in five years and ask me if I thought my internship was a good or a bad experience, I'll probably tell you it was bad but that there were a lot of good things about it. That's what happens to people when they stop being so depraved. Right now, I can assure you there is absolutely nothing good about internship. Nothing.

Well, that's not exactly true. I've worked with a whole bunch of nice people whom I never would have come to know had I not been here. And I had a lot of good times. And I had two wonderful vacations I'll remember for the rest of my life.

See, it's been over for only five hours, and already my mind is warping. Do you think there's any hope for me?

Bob

Wednesday, February 25, 1987

About seven months ago, on a sunny Wednesday morning near the end of last June, as Amy, Andy, and Mark were beginning to celebrate the end of their year of internship, I got into my car and drove up to Peter Anderson's house in Westchester County. At about eleven o'clock that morning I found myself sitting on the grass outside Dr. Anderson's front door and asking three scared-to-death interns-to-be what most worried them. My question was met by an intense silence that lasted for what seemed like minutes. Finally, one of the new interns, a guy named Anthony D'Aquila, meekly said, "The thing I'm most worried about is the night call. I just don't think I'm going to be able to survive a whole year of being on call every third night. I can't understand how you can be up all night every third night and still be able to function the next day."

Slowly, the other two interns joined in, agreeing with

Anthony. Then one of the others, a woman named Andrea Zisman, said that she was worried about what internship would do to her social life. She told us that she'd had a steady relationship with a guy for the past three years; he was a lawyer, and she was concerned that the life-style of an intern would completely destroy this long-term relationship.

We spent about an hour talking together in that group. As the time passed and as the list of anxieties I was recording on the piece of paper in front of me grew longer, I could feel at least some of the nervousness, some of the tension, gradually die away. By the time Mike Miller finally came to call us all to lunch, I had the sense that these three had made some progress; they were ready to begin the year.

I saved the anxiety list from that morning's group and brought it home with me. I compared it with the list that Andy, Mark, and Amy had generated exactly one year before. The lists were almost identical. Not in the same order, not in the same words, but the concerns, the issues, the worries all are universal. Although Andy, Mark, and Amy have moved on, we attendings are dealing with the same problems, counseling away the same anxieties, coping with the same fears in a new group of interns.

Nineteen months have passed now since the day in June 1985 when I asked Mark, Andy, and Amy if they would like to participate in this project by keeping a diary of their internship year. Today those three interns are more than halfway through their junior residency and more than 50 percent finished with their mandatory three-year period of training. And even those three interns with whom I sat at orientation in June 1986 have only four months left until they say farewell to internship. Like the interns who came before them and like the interns who will follow them, they're at present trapped in the depths of the February depression. But I've told them to take heart. The light for them is beginning to appear at the end of the tunnel.

A lot has happened to the public's conception of internship over the past year. Various cases of suspected medical malpractice caused in at least some small part by the fact that unsupervised, overtired, and overwhelmed interns had allegedly made errors in judgment at critical junctures in the management of patients have received a great deal of publicity. The effect of this media attention has been that the lay public's eyes finally have been forced open to the fact that young doctors are often required to work over a hundred hours a week in a system that's antiquated, unnatural, and unhealthy for both the patients and the physicians themselves.

The state of New York has looked into the issue of internship training. The New York State legislature has proposed placing limits on the number of hours a house officer is allowed to work. There are two mechanisms for doing this that are currently being studied. The first of these limits the total number of hours a house officer can work in a single week to eighty; the second limits the number of hours that a physician can work in a single day to twenty-four.

Limiting the number of hours that can be worked in a single day to twenty-four would mean that interns would never again have to work thirty-six-hour shifts; overnight call would be illegal. However, because the wards and emergency rooms have to be staffed twenty-four hours a day, additional house officers would have to be hired. Although the state of New York has announced that funds for these new physicians would be forthcoming, there is a real question whether adequate numbers of medical school graduates could be found to fill these new slots. If sufficient personnel could not be recruited, the new regulations would ensure that interns would wind up working six or possibly even seven days a week. In discussing this possibility with our house officers, almost all state that they would much rather work thirty-six hours at a stretch knowing that they'll have a day off rather than work shorter hours without a day away from the hospital.

But limiting the number of hours an intern can work to eighty a week seems like a viable option. This would essentially outlaw the every-third-night call schedule, replacing it with a more human every-fourth-night scheme.

These reforms are long overdue. But the changes will take time to establish and to institute. So for the time being, at least, internship and residency proceed as they always have.

Early on the morning of June 29, 1986, I drove to the Bronx and stood out in a drizzle as two enormous guys loaded all of Andy Baron's possessions into an Avis rental truck. Andy, Karen, and I stood out there during the hour it took to get everything loaded, getting soaked by the rain. We didn't say much to each other. I knew there were a lot of things going on inside Andy's head, but apparently he didn't want to let either Karen or me in on them. So we just stood there, getting wetter, and silently watched Andy's furniture disappear into the truck.

The two guys were finished by about ten. Andy paid them, and then he and Karen got ready to climb into the truck. Before he got behind the wheel, I put out my hand. Still in silence, Andy came toward me and gave me a bear hug. After a few seconds he released his grip, turned, and climbed into the cab of the truck. A minute later I watched as the truck disappeared down Gun Hill Road, heading east toward the entrance to I-95 and his junior residency at Children's Hospital in Boston.

I've kept up with Andy since that day, speaking with him and Karen by phone a couple of times a month. In some ways, this year has been like a second internship for Andy: He's had to prove himself all over again, he's had to make a whole new group of friends and learn the ins and outs of an entirely new system. Everything about his new program is different from Schweitzer: The patients are mostly private and are referred to the hospital because of the special expertise of members of the faculty. The diseases they have are, for the most part, less common. ("We've got zebras here," Andy told me. "No horses, just zebras.") And the ancillary services are worlds better than ours. It took some period of adjustment, but now Andy is feeling comfortable. Before he left, he had some concern that he would be ill-equipped to work at Children's, that his knowledge and skills after a year of training here in the Bronx would leave him wanting when compared with those of the house officers in Boston. He's told me that that fear has turned out to be unfounded. He feels that he knows as much as if not more than the other junior residents in the program.

Recently, most of Andy's thoughts have been taken up with the future. Specifically, he can't figure out what the hell he's going to do after he finishes his senior year. He's jumped from wanting more than anything to get some subspecialty training so he can know a great deal about one particular area of medicine that very few other people know about, to wanting to be a good primary-care pediatrician, serving the needs of a large number of children and their families. In our most recent conversations, Andy's been leaning back toward specializing. His current favorite area is nephrology. Maybe someday he'll come back to Schweitzer to take care of all the kids at the University Hospital with chronic renal failure.

While in deep sleep during the early-morning hours of Wednesday, November 12, 1986, after nearly two weeks of maternity leave, Amy Horowitz spontaneously ruptured her amniotic membranes and immediately went into labor. Larry, awakened by the rush of warm amniotic fluid that engulfed the bed, immediately jumped up and started to get dressed. They briskly walked the two blocks from their apartment to University Hospital, stopping a few times along the way when the contractions came. Amy was admitted to the labor and delivery suite in active labor. She delivered her second child, a perfectly formed, beautiful boy, just before eight o'clock in the morning. Amy and her son, who was named Eric, stayed in the hospital for three days and were then discharged to home, to spend the next six weeks together until Amy had to return to work.

I spoke with Amy last week. She told me that she can't believe how quiet and well-behaved Eric is. Apparently he never fusses, he rarely cries, and he demands almost no attention. A typical second child! Amy also told me that Sarah loves her baby brother and wants to help with his care whenever possible. "Her biggest goal in life right now is to carry the baby around," Amy said. "But since she weighs about twenty-five pounds and Eric already weighs about twelve pounds, it doesn't look like that's ever going to be possible."

Since her return from maternity leave, Amy has worked very hard and seems to have a serious, no-nonsense attitude about her responsibilities. And her reputation has changed with this apparent change in attitude. I was talking with Eric

Keyes and Enid Bolger, two of the chief residents, about Amy last week. Amy was in the emergency room and had just called up to tell them about an adolescent with DKA [**diabetic ketoacidosis**] whom she apparently had managed superbly. Enid said, "I never worry when Amy's down in the emergency room. She's got a good sense about things. She knows what to do, and when she's in the ER, I know I don't have to worry." Amy's becoming a mother for the second time has apparently caused her to do a great deal of growing.

And what of Mark Greenberg? Of all the people in the internship group, he's probably the one least changed by his transition to junior residency. He's still making everybody laugh. But he has become a leader, which is what a good resident needs to be.

There is another thing that has changed in Mark's life. In July, a few of the house officers were invited to Mike Miller's summer house on Candlewood Lake in Connecticut. Because of Mark's childhood friendship with Mike, he and Carole were invited to come. In the afternoon, Carole and Mark got into Mike's rowboat and rowed out into the middle of the lake. When they stopped, Mark reached into his pocket and pulled out a jewelry box. From the box, he produced a ring. And then he asked Carole to marry him. Old cynical Mark, proposing marriage in probably the most romantic way possible. Carole accepted the ring. They plan to be married this coming summer.

What is there left to say? My own internship was the hardest, most devastating year of my life. It's been eight and a half years since I finished that year, and some of the pain, the anger, the exhaustion, and the anguish is still with me. I don't think my experience, or the experiences of Andy, Amy, and Mark, are unique. Everybody who lives through an internship is forever changed by the experience. The intern learns about medicine and the human body; he or she truly becomes a physician. But in the process, through the wearing down of the intern's spirit, that person also loses something he or she has carried, some innocence, some humanness, some fundamental respect. The question is, Is it all worth it?

AFTERWORD

January, 2001

"Is it all worth it?" A little less than fifteen years have passed since I wrote those words ending the manuscript of *The Intern Blues*. Even though by that time nearly a year had passed since Andy, Mark, and Amy had finished their internships, each was still engrossed in residency training; the experience was still too close, both temporally and emotionally, for them to be able to offer a valid response to my question. But now, with the perspective of time, of more than ten years since the completion of their training, they should be able to look back and offer some insight. So, when my friends at HarperCollins informed me that they wanted to put out a new edition of *The Intern Blues*, I began to search for the three ex-interns. In addition to getting an answer to the question I'd posed a decade and a half ago, I wanted to see what had happened in their lives since then.

* * *

Near the end of the epilogue to *The Intern Blues*, I stood in the rain, watching the truck Andy Baron had rented disappear down Gun Hill Road. With Andy driving and his girlfriend, Karen, sitting in the truck's passenger seat, they were hauling Andy's stuff from his apartment in the Bronx to their new place near Children's Hospital in Boston, the facility at which Andy would soon begin his junior residency year. Although I did keep in contact with Andy via periodic telephone calls for the first year or so, we lost contact soon after and had not spoken since.

Finding Andy was quite easy: I simply searched for him on yahoo.com. Typing his name into the search engine, I immediately found him listed among the faculty of the Division of Neonatology on the website of the Department of Pediatrics at Boston University School of Medicine. After calling the main office, I was given the number of the Neonatal Intensive Care Unit at Wellesley Medical Center, a community hospital affiliated with Boston University, in which— at least according to the secretary—Andy worked. I dialed the number, and within a minute, he was on the line. To say the least, Andy was surprised to hear from me.

We spent a few minutes catching up. After driving that truck down Gun Hill Road, Andy got on I-95 North and headed for Boston, where, as expected, he spent two years completing his residency at Boston Children's. Then, feeling wiped after the long, arduous process of residency training, he took a job working in a local pediatric practice, seeing patients three days a week. "After my residency, I felt . . . I don't know . . . 'used up' is probably the best way to describe it," he told me. "I needed to replenish myself. I didn't have any direction, I didn't really have a plan about what I wanted to do with the rest of my life. I knew I didn't want to do general pediatrics forever, but the job gave me a chance to clear my head and think about the future."

Andy got married during this period—to Karen, who was then doing her residency training in psychiatry—and they remain married ("happily married" in Andy's words) to this day.

The general pediatric practice eventually grew to bore Andy. He worked there for a total of two years, then took another part-time job, this one doing shift work as a hospital-ist [**essentially a glorified resident**] in a neonatal intensive care unit. "I always liked working in the NICU," he told me. "There was always something exciting going on. I even liked it during residency, but back then it was mostly unpleasant because of the hours they forced us to put in. But when you're working shifts, you can decide how much you want to work and how much you want to screw around. I played with my schedule until I had what I thought was the perfect mix."

Andy liked working in the NICU so much (and was so good at it) that he decided to obtain formal training in the specialty, a step that would ultimately allow him to become a board-certified neonatologist. So four years after completing his residency, he returned to Boston Children's as a neonatal fellow. "Those were three pretty grueling years," he explained. "Each of my fellowship years was easily as hard as internship."

Following completion of his fellowship, Andy passed the boards in neonatology and took the job he currently holds, director of the nursery at Wellesley Medical Center. "It's a small unit. We have only eight beds. Most of the time it's manageable. But on those days when we're really busy, it's hard to get out of here at night." He also continues to work only half-time, splitting coverage with two other part-time neonatologists. "When Karen and I first started thinking about having kids, we decided we didn't want them to be cared for by nannies or grandparents or anyone other than ourselves. So we agreed that we'd each work part-time and spend the rest of the time caring for the kids." The Barons have two boys, currently seven and three years old. "It's worked out really well for us. Don't get me wrong, it hasn't always been easy. We've had to make some sacrifices. Believe me, we could be making a lot more money if we both worked full-time. But we're comfortable, and how much money do we need? This way, Karen and I both have lives. I have time to paint and draw, things I used to love to do during high school and college but never had time for once I started medical school. And I'm able to watch my kids grow up. That's worth a lot more to me than having a ton of money."

I asked Andy to look back and consider what effect his internship had on him. He thought about the question for a while before answering. "You change so much during your training," he replied. "You see such awful stuff, it has to have a permanent effect on the way you look at the world. But you change a lot as you get older, anyway, whether you do an internship or not. So I don't think I can blame internship or residency for any big epiphany that occurred to me.

"The biggest change in my life happened when my first son was born," he added. "Having kids definitely changes you; they change your perspective on patients, on their families, everything. I began to talk to my patients' parents a lot differently after having my own child. I had a better understanding of what was important to them, and what wasn't so important. Having a child definitely had more of an effect on how I deal with patients than anything that happened during my internship."

And was it all worth it? Again, Andy considered the question for a few seconds before answering. "I'd say yes," he finally replied. "I like what I do; I like being a neonatologist, and being able to set my hours and live the kind of life I live. None of this would have been possible had I not done an internship. There might be better ways of doing it, better methods of training young doctors, but all in all, it was a means to an end. I made it through and here I am. But there's another question you should ask me, Bob."

"What's that?"

"Would I want my sons to become doctors?"

"Well, would you?"

"No. But it has nothing to do with my training. It has to do with insurance. The insurance industry has made medicine completely crazy. If you had told me when I was an intern about what I'd wind up having to do to get insurance companies to pay for what I think my patients need, I'd have told you that you were nuts. HMOs and the rest of the insurance industry have made the practice of medicine horrible, and I wouldn't want my boys to have to go through this."

We talked for a little while longer, but Andy had to get back to work. I wished him well, told him to give Karen and his kids a kiss for me, and we hung up.

Mark Greenberg was a little harder to locate: In contrast to Andy, he does not have a presence on the Internet. In order to find him, I sniffed around our department's alumni records and came upon a letter of recommendation written by Peter Anderson (our chairman and the man whose lawn was the site of the original meeting between Andy, Mark, Amy, and me) and addressed to a hospital in central New Jersey to which Mark had applied for admitting privileges. Calling information in that town, I was given Mark's office number. I hesitated before dialing the number.

I hesitated because Mark and I had not parted on exactly the best of terms. It took me more than a year to transcribe and edit the audio-diaries of the three interns that formed the basis of *The Intern Blues*. As I completed that work, I sent each of the by-then senior residents a copy of his or her transcript. A few days after sending his out, I got a call from Mark. "You can't publish this crap," he yelled at me over the phone. "I never said any of this stuff."

"What stuff are you talking about?" I asked, a little surprised by his reaction.

"All this stuff," Mark continued yelling. "Almost everything you have coming out of my mouth. Like here, during February, when I was in the NICU. You wrote, 'And then we walked around and he showed us these so-called patients. My God, those things weren't patients; they couldn't have been human; they weren't anything more than small pockets of pus and protoplasm! These things would have to quadruple their weight in order to be classified as patients. Right now, they're nothing more than tiny portions of buzzard food.' Bob, I never would have referred to preemies as 'small packets of pus and protoplasm' or 'tiny portions of buzzard food!' "

"You don't remember saying those thing?" I asked, surprised.

"I never said them!" he answered. "How could I remember saying them if I never said them?"

"You don't remember your internship very well, do you Mark?"

"I remember it fine," he replied. "I agree that things were rough for most of the year. But no matter how bad things got, I always showed respect for my patients. I'm sure of that."

"Mark, I hate to burst your bubble, but not only did you say those things, I've got you saying them on tape. Would you like me to play them back to you?"

"You can't have them on tape, because I never said them," he reiterated, more angry this time. "If you have someone saying that stuff on tape, then it must have been one of the other interns, because it couldn't have possibly been me."

We argued on like that for a while. Finally, I agreed to let Mark "fix" at least some of what he believed was wrong with his portion of the book. I used his edited transcript to revise the final manuscript, but he still wasn't happy with the finished product. In the final eighteen months of his training, that unhappiness colored our relationship. So I was less than sure that Mark would react positively to my voice on the phone.

I called his office at about eleven o'clock on a Monday morning in late January, and when he picked up the phone, I was afraid my worst fears would be realized. "I can't talk now," he said bluntly when I told him who it was and why I was calling. "It's crazy here. Can I call you back sometime after our office closes?" I gave Mark my home number, never expecting to hear from him again.

But I was wrong; he called me that night. "Sorry about not being able to talk with you earlier in the day, but it's been really nuts," he explained, sounding neither angry nor put-out. "It's flu season here in beautiful New Jersey, and it appears as if everyone with a pulse and a respiratory rate is sick. I was on call this weekend, and I saw seventy patients on Saturday and fifty more on Sunday. A hundred and twenty patients in one weekend! By myself! Can you believe that? It's a new office record. In recognition, I'm planning on having the shoes I was wearing bronzed so we can display them in our practice's trophy case. So, sorry I blew you off this morning, but I think it'll work out better this way."

"I thought you were still pissed off about the way you were portrayed in the book," I said.

"Was I pissed off about that?" he asked. "I don't remember. I really don't remember much about my residency. Except that I hated it as much as anybody can hate anything, and that I'd never want to do it again. Outside of that, I don't remember the people, the places, the patients, or much of anything else."

"You don't remember making me rewrite your sections?" I asked.

"The only thing I remember about it is I was afraid people would recognize me and give me a hard time. Remember, I had to go out and look for a job. I thought people in practice were all going to know I was Mark Greenberg in *The Intern Blues*, and who would want to hire someone who talked about patients the way that guy did? As it turned out, it wasn't an issue; I don't think anyone had even heard of the book. Looking back, I might have been a little crazy at the time."

I silently agreed. "So what have you been doing since the late eighties?" I asked.

"Well, as soon as I finished residency, I started working at this practice," he replied. "I joined in July of 1988, and I've been working here ever since. I've been here for nearly thirteen years, I've been a partner for ten, and I only have eleven more years until I retire."

"Until you retire?" I repeated. "You've thought about when you're going to retire?"

"Sure I know when I'm going to retire. I plan everything. When I started here, I told them I was going to work until I turned 55. The group was fine with that; they all agree with it."

"Why 55?" I asked.

"You've obviously never been in practice," he replied. "You can't do this kind of work forever. Did you hear what I said before? I saw one hundred and twenty patients over the weekend. One hundred and twenty! By myself! And that doesn't count doing rounds at the hospital and speaking on the phone to the mother of every one of our patients who didn't come into the office. You can't do all that—the hospital, the office visits, the phone calls night and day—when you're 60 years old. Pediatric practice is for young people. And so I'm going to stop when I turn 55."

From what Mark was saying and the way he was saying it, I got the impression that he didn't enjoy his job all that much. "Do you like what you're doing?" I asked.

"I love it," he answered without hesitation. "I'm doing exactly what I wanted to do from the time I started medical school. Of course, back then, I thought I was going to do adult medicine. It wasn't until my third-year rotations that I realized I liked kids and hated adults. But basically, yes, this is what I thought I'd be doing, having a practice, seeing patients. I get a real thrill out of seeing patients. There's always something new. Plus, I'm the doctor for the local high-school track team, and, in the summer, I watch over the local Girl Scout camp. It's a lot of fun."

Next I asked Mark about his life away from the office. I reminded him that in the epilogue to *The Intern Blues*, I had described how he had romantically proposed to his then-girlfriend, Carole, in a rowboat on Candlewood Lake in Connecticut. "Did you two wind up getting married?"

"We certainly did get married," he replied, "but not in a rowboat. We got married in a hotel ballroom in Manhattan. We had been living together since we got engaged, and I didn't think it would be right for me to join the practice without us being married. I mean, what kind of an example would I be setting for my patients? So we tied the knot in May of my senior residency year. Carole wasn't all that anxious to go ahead with it just then. She was in law school and didn't have time for anything but studying. But I insisted and told her I'd take care of everything. I planned the whole thing, from the invitations to the honeymoon. All Carole had to do was show up. Afterwards, she told me I did such a good job that if my job didn't work out, I could always become a wedding planner."

"Do you have any kids?"

"One. A boy named Alex. He's eight."

Remembering what Andy had said about the ways that the experience of being a father had affected his work, I asked if Mark had had a similar experience. "Not really," he explained. "I don't think I approached my work any differently before Alex was born than I do now."

Also, as with Andy, I asked Mark to think back and consider what effect internship had had on him. "As I said before, Bob," he answered, "I don't remember much about my internship, except that it was bad, definitely the worst year of my life. Nothing comes close to how bad it was. And the worst part wasn't the hours or the sleeplessness, or anything like that, although those things were pretty bad. The worst part was the frustration and the fear that you were going to make a terrible mistake and kill some kid. They were taking people who hadn't been trained to do things, who never had experience caring for critically ill patients on their own, and forcing them into situations where they were the ones making the decisions. That wasn't fair to the interns, and it especially wasn't fair to the patients. I think they've done a lot to fix that; I think the new laws that require that interns and residents be better supervised are definitely a good thing. But that wasn't the case when I was an intern. And because of that, it was by far the worst experience of my life."

"Was it worth it?"

Mark's response was essentially the same as Andy's: "Yeah, I think it was worth it. It's like my running. Since I was in high school, I've been a marathon runner. In order to qualify to run in a marathon, you have to prove you're worthy. In other words, you have to do a lot of shit in order to get to where you want to be. Internship is the shit you have to do in order to practice medicine. I like my job; I like my life. And doing internship and residency was the only way I could get to this point. So definitely, yes, it was worth it. Just don't tell me I have to do it over again. Boy, that would be bad!"

When she left the Bronx, Amy Horowitz left virtually no footprints. After all these years, she was by far the most difficult of the three former interns to locate. There were no hits on any Internet search engines, no letters of recommendation, no requests for verification of training. No other information on her whereabouts was revealed during a thorough search through her files in our departmental offices. I even called the

Office of Alumni Affairs at the Albert Schweitzer School of
Medicine for help; although they had her name listed as a
bona fide graduate of the school, they had her listed as "lost,"
with no current address available. It looked as if I was stuck.

I was about to give up when I decided to try one last
strategy. Figuring that her father would know where she was
(Amy's mother had passed away while she'd been in college),
I called Schweitzer's admissions office, hoping to get the
home address she'd listed when she'd filled out her medical
school application twenty years ago. A secretary in the office
volunteered to try to retrieve Amy's application from its
microfiche resting place; successful, she called me back in less
than an hour with both the phone number and address. Calling the number, I didn't have a great deal of hope that Amy's
father would still live there; still, I felt I had to give it a shot.

It turned out that I was correct: Amy's father no longer
lived in New Jersey. The woman who answered the phone,
though, had first-hand knowledge of where he was. She
turned out to be his niece, the daughter of Amy's father's sister, and she and her family had been living in the Horowitz
house for the past seven years. Her uncle, she told me, had
retired and moved to Boca Raton, Florida. "And Amy's been
living in Israel for more than ten years," she said. She gave me
her uncle's number, and after a few minutes on the phone
with him, explaining who I was and why I wanted to get in
touch with his daughter, I finally got the information I'd
been searching for. So early the next day, I called Israel. The
phone rang three times before Amy picked it up.

She and I had last spoken in 1989, and sadness had
infused that discussion. In the epilogue of *The Intern Blues*, I
wrote about the birth of Amy's second child, a son named
Eric, who had been born on November 12, 1986. After the

two months of maternity leave and one month of an elective without night call, for which she'd fought so hard, she returned to residency full-time. In the last eighteen months of her training, Amy really put herself into her job. She worked hard and became a leader among the residents, and she had a good sense of what needed to be done for her patients and the ability to go ahead and do it.

In the late winter of her senior residency year, Amy had become pregnant again. This time, she'd planned the pregnancy to the last detail, being due to deliver three months after her training was scheduled finally to come to its end. "I've lived through a pregnancy as a resident," she had told me when she called. "I never want to have to do that again." She would finish her residency, spend the last trimester of her pregnancy relaxing, give birth, and spend the rest of that academic year being a mother to her three children. "We'll be like a real family for a while. Then, depending on how things work out, I'll either go back to work or I'll spend another year as a mommy. We're flexible."

Right on schedule, on the morning of October 18, 1988, Amy delivered a beautiful baby boy. But within twelve hours, her family's world was turned upside down. The baby developed perioral and acral cyanosis [**blueness around the mouth, hands, and feet, often caused by heart disease**]. The pediatric resident on call that night at University Hospital did an arterial blood gas, which showed an oxygen level of only fifty-eight in room air (a normal value would have been in the nineties). When the baby was placed in an environment of 100 percent oxygen, the level increased only to seventy-two (under normal circumstances, the oxygen level in the blood should have risen to more than 200). An emergency cardiology consult was requested. The cardiologist, who made it to the hospital at around 2 A.M., did an echocardiogram, which revealed the terrible problem: Amy's baby had a hypoplastic left heart.

Hypoplastic left heart is a catastrophic congenital defect in which, for reasons that are never clear, the left ventricle, the chamber of the heart that pumps oxygenated blood to the rest of the body, doesn't develop. During fetal life, the defect doesn't create much of a problem. The fetus has a structure called the ductus arteriosus that shunts blood that would normally go to the lungs to the remainder of the body (fetuses don't need to use their lungs; they receive oxygenated blood from their mothers through the umbilical cord). Soon after birth, in a transition from fetal to adult circulation, the ductus arteriosus closes, allowing blood to flow through the pulmonary artery to the lungs. In most babies, this change facilitates normal babyhood; in Amy Horowitz's baby, however, the closure of the ductus arteriosus was a death sentence.

Amy knew all this. As soon as the cardiologist came into her hospital room and told her what was wrong, she understood completely the implications of the news. She knew that if left untreated, her son would be dead within thirty-six hours. She also knew that there was an option: Surgery could be done essentially to build the baby a left ventricle. The complete surgical repair would involve multiple stages. An initial operation would need to be done as soon as possible; a second operation would have to be performed later in the first year of life; and at least one other operation would be required when the baby was older. And this was not simple surgery: Each operation was fraught with risks. The first one alone had a 50 percent mortality rate. And even in the best case, even if her baby did manage to survive each of the stages of the complex repair, he would have a terrible first few years of life and would face an uncertain future.

In addition to all of this, Amy had to be concerned about the rest of her family. There were her other two children to think about. Surgery would mean an all-out commitment to this baby, who would spend the better part of his first year of life in the hospital, much of that time in the intensive care unit. Would it be fair to Sarah and Eric to deprive them of the love and attention of their parents during long stretches of time in order to provide for the new baby, whose future, under the best of circumstances, was far from certain?

On the other hand, Amy thought, this was her baby, her son. He had grown and developed in her womb for nine months. Knowing that a treatment was available (even though the treatment was not even close to guaranteed to work) and that that treatment might give him a life, could she deprive him of this chance simply because life would be hard for the rest of the family? She was not sure.

The cardiologist understood her concerns. He started the baby on a Prostaglandin drip, an intravenous medication that would keep the ductus arteriosus open while Amy and Larry made a decision about what to do. Later that morning, from her hospital bed, Amy began calling everyone in the world she trusted and respected, telling her terrible story and asking for advice. I was on the list of people she called. During our conversation, I mostly listened to what she had to say. It seemed clear from her words that she had already made up her mind; she was looking for approval for her decision to stop the Prostaglandin and let nature take its course. I did as she implicitly requested: I agreed with her reasoning and tried my best to offer support.

By late that afternoon, the decision was finalized. She and Larry met with the neonatologist and told her that they wanted the Prostaglandin drip turned off. After the IV was pulled, the baby was brought to Amy's room on the maternity floor. He died in his mother's arms early the next morning.

In the months after the death of her son, I spoke with Amy every few weeks. On October 20, the family held a private funeral. Not unexpectedly, following this, Amy went through a period of depression. But the demands on her time, caused by the need to care for Sarah, who was then three and a half years old, and Eric, who was nearly two, prevented her from becoming dysfunctional. "I have to get up every morning and get Sarah ready for nursery school," she told me during one of these conversations. "And Eric needs everything done. The only time I get to myself is when he takes his nap. For me, that's the worst time of the day. That's the time I sit and think about exactly what I did."

In early spring of 1989, Amy told me that she and Larry had decided to move to Israel. "There's not a lot for us here," she explained. "I don't have a job and Larry's not happy with his. The only family I've got is my father, and he's thinking of retiring and moving away. And there are all these memories of what's happened."

I hadn't spoken with her since that day. But now, here she was, picking up the phone after the third ring and saying "Shalom."

She said she was happy to hear from me. "We live just outside of Jerusalem," she told me, sounding exactly as she had more than ten years before. "Larry, me, and our four kids."

"Four kids!" I repeated. "When you left New York, you only had two."

"That's right," she answered. "Our last two are sabras. Aviva was born in 1990, and Seth, our baby, was born in 1992.

"And Sarah and Eric must be teenagers."

"Right, again. Can you believe it? Sarah's going to be sixteen in a couple of months. And Eric's fourteen. It's been great watching them grow up."

"Have you been working?" I asked.

"Well, I've been a mother, which is a full-time job. But no, since coming to Israel, I haven't worked as a pediatrician. That's not the way I planned it; when we first arrived, I thought I would. I investigated how to go about getting licensed. But it was a lot of trouble. At the time, there was a glut of doctors here, and they were making it as difficult as possible for physicians who had trained outside the country to get jobs. To me it was no big deal: I had the kids to take care of, Larry had a good job, we didn't have to worry about money. Since then, the time's just passed, I've taken care of my children, and here we are in 2001."

"Do you miss medicine?"

"No, not at all. When I started medical school, I really did want to be a pediatrician. But somewhere along the way, I just stopped wanting it so much. Residency kind of beat the desire out of me. And then when my baby died . . . well, that kind of put the icing on the cake. After living through that experience, I just didn't care about medicine anymore. It was just a job, something to do to earn money, and since we didn't need money, there was no need for me to do it."

Knowing the answer in advance, I asked whether she thought her internship was worth it. "Internship was the second-worst experience of my life," she responded. "I don't have to tell you what the first-worst experience was, because you were around for that, too. But other than watching your baby die, I can think of nothing I'd want to do less than spending a year as an intern. It was degrading, depressing, and frightening. It brought out the worst in me, in my class-mates, in those in charge of the program, in everyone. It was unhealthy, and it was just unfair—unfair to us, unfair to the patients, unfair to society. Back when I was doing it, I couldn't see why I had to do it, and looking back on it now, I still can't understand why I had to do it. So Bob, you can put me down for a 'no' vote. I don't think it was worth it!"

And so, as in so many different events that occurred in this book, we're left with a difference of opinion. Amy Horowitz, who following completion of her training chose not to prac-tice medicine, believes that internship could never be worth the torture she was forced to endure. Andy Baron and Mark Greenberg, however, agree that internship was worth it, because it proved to be a means to an end, the end being a career in medicine. They never would want to go through it again, though.

Something Andy told me during our recent conversation is still reverberating in my head. "Internship is like a bad dream," he said. "When you're having it, it seems like the most terrifying thing in the world. In the morning, it still seems pretty bad, but in the light of day, the details are blurry and don't make a lot of sense. Then, after a few days have passed, you almost can't remember it anymore. You know you had a dream, you know it was scary, but all the details are gone."

I think that's how most physicians remember their internships.

abruptio placentae Separation of the placenta from its attachment on the wall of the uterus. Dangerous because it deprives the fetus of oxygen.

acidosis Buildup of acid in the blood.

acute abdomen Term used to describe condition in which an abdominal catastrophe is occurring. Causes of acute abdomen include ruptured appendix, and inflammation of the pancreas and gallbladder.

agonal rhythm A pattern of electric activity generated by the heart just prior to death.

A-line Abbreviation for arterial line, a tube placed in an artery used for sampling blood.

ALL Abbreviation for acute lymphocytic leukemia, a common form of childhood cancer.

AMA Abbreviation for "against medical advice," the situation in which a patient signs out of the hospital before his doctor considers it safe to do so.

ambubag A device used to force air into the lungs; consists of a mask that covers the mouth and nose and a rubber bag that, when squeezed, generates a gust of air under pressure. Used in respiratory arrests.

aminophylline Drug used for treating asthma.

aneurysm of the ascending aorta A ballooning out of the aorta, the main artery bringing blood to the baby.

antecubital vein A vein at the elbow that usually is most accessible for blood drawing.

APGAR scores A scoring system, designed for use in the newborn, that monitors fetal and neonatal well-being. Maximum score is 10.

ARC Abbreviation for AIDS-related complex.

ARDS Abbreviation for adult respiratory distress syndrome, a condition in which respiratory failure occurs.

argininosuccinicaduria A rare inborn error of metabolism that leads to mental retardation and often to premature death.

aspiration The act of breathing in; aspiration pneumonia results from the breathing in of foreign substances, such as meconium in the newborn.

atelectasis Collapse of the lung.

ATG Abbreviation for antithymocyte globulin, a drug used in patients who are rejecting transplanted organs.

bagging The act of forcing air into the lungs of a patient having difficulty breathing.

BCW Bureau of Child Welfare, the New York State agency charged with investigating and acting upon cases of suspected child abuse.

Benadryl An antihistamine used in treatment of allergic reactions.

beta HCG A blood test to diagnose pregnancy; more accurate than a urine pregnancy test.

bicarbonate A drug used in patients with acidosis (*see above*).

bladder tap Procedure in which a needle is passed through the lower abdominal wall and into the bladder so that an uncontaminated sample of urine can be obtained.

blood gas A test done on a sample of arterial blood that tells the amount of oxygen and carbon dioxide within the body.

bradycardia Abnormal slowing of the heart rate.

bronchiolitis Inflammation of the bronchioles, the small air passages leading to the lungs. This is a condition that resembles asthma and occurs in children under a year of age.

bronchopleural fistula A leak in the main air tube that causes air to leak into the pleural cavity, causing a tension pneumothorax.

BUN Abbreviation for blood urea nitrogen, a substance that builds up in renal failure and in dehydration.

CAC Abbreviation for "clear all corridors"; a general request, announced over the loudspeaker, for help at a cardiac arrest.

cadaveric transplant A transplant performed using an organ obtained from a person who has died.

cardiac cath A procedure in which a catheter is placed in an artery or vein and threaded up to the heart, at which point a dye is inserted. Used to define the nature of heart disease.

CBC Complete blood count. A blood test to examine the content of hemoglobin, red blood cells, and white blood cells within a sample of blood.

central line A tube placed in one of the major blood vessels, usually in the neck or the groin.

cervical adenitis Swelling of the lymph nodes of the neck.

chemo Short for chemotherapy.

chief of service rounds A weekly conference in which an

interesting case is presented and discussed by an expert in the field.

choriocarcinoma A cancer that develops from a molar pregnancy.

CIR Committee on Interns and Residents, the house staff union.

CP Cerebral palsy, an abnormality usually caused by lack of oxygen around the time of delivery.

CPR Cardiopulmonary resuscitation; technique used at cardiac arrests to keep the patient alive.

cracking the chest The surgical opening of the chest wall to gain access to the heart and lungs.

creatinine A substance in the blood that's elevated in cases of kidney failure.

crump To deteriorate rapidly.

CSF Cerebrospinal fluid, the liquid that bathes and protects the brain.

CT scan An X-ray procedure; most CT scans examine the head, looking for specific defects of the brain.

cutdown A surgical procedure in which a vein is isolated and a tube is placed into it for access. Used when no superficial veins can be found.

CVA tenderness Tenderness at the costrovertebral angle, the area of the back under which the kidneys are situated. CVA tenderness is present with many types of renal disease.

cyanosis Blue discoloration of the skin usually due to lack of oxygen in the blood.

dactylitis Swelling of the hands and feet; usually the first presentation of sickle-cell disease in an infant.

DKA Diabetic ketoacidosis; a severe metabolic abnormality that occurs in diabetics who have a marked buildup of sugar in their blood. If not cared for correctly, it may lead to death or brain damage.

DNR Do not resuscitate.

double footling breech Condition in which baby is heading out of the birth canal feet first; dangerous because there's a chance the fetal head can get caught, leading to the baby being unable to be born.

DPT Two meanings: (1) immunization given to young children that protects against diphtheria, pertussis (whooping cough), and tetanus; (2) a mixture of Demerol, Phenergan, and Thorazine used to sedate children who are having some painful procedure.

dual response Procedure used in cases of child abuse in which the child is considered to be in danger. The police and the BCW (*see above*) are informed, and an immediate investigation is carried out.

dysplastic kidneys Condition in which the kidneys did not form normally and therefore cannot function well. Often leads to chronic renal failure.

E. coli A type of bacterium.

ectopic pregnancy A pregnancy occurring in a location other than the womb.

EEG Electroencephalogram; a test in which the electric activity of the brain is examined.

EMS Emergency Medical Service; the agency that staffs the ambulances.

endotracheal tube A tube passed through the larynx and into the main breathing tube that allows the individual to be placed on a respirator.

epinephrine A drug used in patients whose hearts have stopped beating.

ER Emergency room.

extramural delivery Delivery of a baby that occurs outside of a hospital.

fascinoma An interesting case.

febrile seizure A convulsion caused by a marked elevation in fever.

FFP Fresh frozen plasma; a part of the blood that sometimes is used as a transfusion.

FIB Fever in baby; all babies under two months of age with fevers should be admitted to the hospital and treated with antibiotics.

Foley catheter A tube that has an inflatable balloon at the end; Foleys usually are passed into the bladder to help monitor urine production.

fundoplication A surgical procedure used to correct gastroesophageal reflux (GER—*see below*).

gastrostomy A surgical procedure in which a hole is made in the abdominal wall and the stomach so that feeding through a tube can be accomplished without the patient's having to suck and swallow.

GC Gonococcus, the bacterium that causes gonorrhea.

GER Gastroesophageal reflux, the regurgitation of stomach contents back into the esophagus.

gonococcus The bacterium that causes gonorrhea; also known as GC (*see above*).

gram-negative rods The appearance, under the microscope, of certain bacteria when stained with dye using a special technique. *E. coli (see above)* is the most common form of gram-negative rod.

G-tube A tube placed through the opening of a gastrostomy and through which blenderized food is squirted.

guaiac A test looking for blood, usually in a sample of stool.

Haldol A drug used in patients with psychosis.

H and P History and physical; the admission note that must be completed on all patients staying in the hospital overnight.

headbox A Plexiglas box that fits over the head of an infant and through which oxygen in high concentrations can be administered.

hemifacial cellulitis Infection of the skin on one side of the face.

hemoptysis The coughing up of blood.

HIV Human immunodeficiency virus; the agent that causes AIDS.

hydrocephalus Dilatation of the ventricles of the brain, which can lead to increased intracranial pressure.

hyperbaric chamber The center on City Island in New York City where the barometric pressure can be increased. Patients who have inhaled smoke and have a high level of carbon monoxide in their blood are sent there for treatment.

hypoxia A deficiency of oxygen in the blood. May lead to brain damage.

infantile spasms A particularly severe and damaging type of seizure disorder.

interosseous infusion Procedure in which fluids or drugs are injected into the bone; used in dire emergencies when an IV cannot be started.

intracardiac infusion Procedure in which drugs are injected directly into the heart; used as a last-ditch effort to save a patient's life when IV access cannot be established.

intracranial bleed A hemorrhage into the brain; usually causes brain damage or death.

intubation Procedure in which an endotracheal tube (*see above*) is inserted through the vocal cords and into the main breathing tube.

Isuprel drip Constant infusion of isoproterenol, a drug used in severe asthma.

IVDA Abbreviation for intravenous drug abuser.

kerlix Rolls of bandages, often used to restrain little children.

KUB Abbreviation for kidneys, ureter, bladder; a type of X ray in which the abdomen is examined.

Lasix A diuretic drug used in hypertension and heart disease.

lidocaine A local anesthetic.

LMD Abbreviation for local M.D. The patient's private doctor.

LMP Last menstrual period. Important because it's used to determine how premature a baby is.

LP Lumbar puncture; a procedure in which a needle is inserted through the back and into the spinal canal so that a sample of CSF (*see above*) can be obtained for study. Also known as spinal tap.

main-stem bronchus One of the two main tubes connecting the trachea and the lungs.

M and M conference Morbidity and mortality conference, a teaching exercise at which a patient who has died is discussed.

meconium The baby's first bowel movement; when meconium is passed while the baby is still in the womb, it often is a sign of fetal distress and can lead to respiratory problems if it is aspirated.

mediastinum The central part of the chest that houses the heart.

membranes The structures that contain the fetus, the placenta, and the amniotic fluid. Rupturing of the membranes, followed by a gush of amniotic fluid, often causes the onset of labor.

meningomyelocoele A defect in the spine, present at birth, that often is associated with hydrocephalus, neurologic deficits of the legs, and urologic abnormalities. Also called spina bifida.

methotrexate A chemotherapeutic agent used in the treatment of some cancers.

"mets to the brain" Metastatic cancer affecting the brain.

mitral stenosis Tightness of the valve that separates the heart's left atrium and left ventricle.

monilia A type of fungus that frequently causes diaper rash in infants; also affects patients with immune deficiencies such as AIDS.

NEC Abbreviation for necrotizing enterocolitis, a severe disorder affecting the intestines of some premature babies.

nephrotic syndrome A condition affecting the kidneys that results in the inability to retain protein.

neuroblastoma A relatively common form of cancer that affects children.

neurofibromatosis A genetically inherited disorder that can cause abnormalities of the skin, the central nervous system, and other organs. Also known as the Elephant Man Disease.

NICU Abbreviation for neonatal intensive-care unit.

night float A resident who is scheduled to work the overnight shift in the emergency room.

NPO Abbreviation meaning "nothing by mouth." Ordered for patients with intestinal abnormalities and patients who are pre-op.

occipital hematoma Hemorrhage into the back part of the skull or the underlying brain.

OPD Outpatient Department, composed of the ER and clinics.

orthopods Internese for orthopedic surgeons.

osteogenic sarcoma A type of cancer affecting bones.

otitis media Infection of the middle ear. Very common cause of fever in infants and young children.

oxacillin A type of antibiotic.

painful crisis A complication/result of sickle-cell disease; sickling of red blood cells leads to lack of oxygen reaching the tissues and results in development of severe pain.

pancytopenia Deficiency of all types of blood cells, both red and white.

patent foramen ovale An opening between the two atria of the heart. If untreated, it might eventually lead to pulmonary hypertension (*see below*).

Pavulon A drug that paralyzes the recipient; used in patients on respirators who are agitated and said to be "fighting the machine."

PDA Abbreviation for patent ductus arteriosus, a congenital defect of the cardiovascular system that is common in premature infants.

perineum The genital region.

periorbital cellulitis An infection of the skin surrounding the eye. Dangerous because it can lead to infection of the eye (orbital cellulitis), which can lead to infection of the brain.

peritoneal dialysis A procedure performed to "cleanse" the blood in patients with kidney failure, in which fluid is placed into the abdominal cavity and later drained out.

PFC Abbreviation for persistence of fetal circulation, a complex physiological abnormality encountered in newborns who have aspirated meconium.

pH Measure of acid in the blood. Determined routinely as part of a blood gas (*see above*).

phototherapy A treatment for jaundice of the newborn in which the infant is placed under ultraviolet light; through a mechanism that's not clear, this therapy lowers the level of bilirubin in the blood.

physsies Internese for physical examinations of the newborn performed in the well-baby nursery.

PID Abbreviation for pelvic inflammatory disease, an infection of gynecological structures. Often caused by gonococcus (*see above*).

pneumococcal meningitis An infection of the spinal fluid caused by a very virulent and damaging form of bacterium.

pneumothorax Collapse of a lung; must be treated by placement of a chest tube that drains out the accumulated air.

PRN Abbreviation used in medication orders meaning "as needed."

prolapsed cord Condition in which the umbilical cord passes out through the cervix before the baby. Dangerous because if the cervix narrows, blood flow to the fetus can be blocked, leading to hypoxia (*see above*) and brain damage.

pseudomonas A virulent bacterium.

P²C² Abbreviation for Pediatric Primary Care Center, the clinic at Jonas Bronck Hospital.

pulmonary hypertension Increased blood pressure in the pulmonary arteries, usually as a result of heart disease. Is a nonreversible condition that eventually will lead to death.

pyloric stenosis Narrowing of the lower part of the stomach, leading to inability to pass stomach contents into the intestine. Occurs most commonly in first-born male infants during the first two months of life.

q4h When written in a medication order, means "every four hours."

QNS Abbreviation for quantity not sufficient.

rales A particular sound heard when listening to the lungs through a stethoscope and that implies the presence of pneumonia.

renal biopsy A technique in which a needle is passed through the back or side and into the kidney. Allows sampling of kidney tissue and therefore diagnosis of specific diseases affecting the kidney.

'roids Internese for steroids, an anti-inflammatory class of drugs.

scut A collective term for the routine work that an intern must do.

sed rate Short for erythrocyte sedimentation rate, a test used to determine if an inflammatory disease is occurring.

sepsis Bacterial infection in the blood.

SGA Abbreviation for small for gestational age. Used for babies who have not grown to adequate weight while in the womb.

sickle-cell anemia An inherited disorder in which an abnormality of hemoglobin, the protein that carries oxygen, causes deformation of the red blood cells and serious consequences. (*See* painful crisis, *above.*)

sigmoidoscopy Examination of a portion of the colon using a device called an endoscope.

spina bifida Synonym for meningomyelocoele (*see above*).

status epilepticus Condition in which patient is having constant, uncontrolled convulsions.

straight cath Passage of a tube through the urethra into the bladder to obtain a sterile sample of urine.

strawberry hemangioma A purplish birthmark; these often disappear by the time the child is six years old.

subarachnoid hematoma A collection of blood between the arachnoid membrane and the brain.

subinternship A two-month rotation in the fourth year of medical school in which the student acts as an intern.

suprapubic tap Synonym for bladder tap (*see above*).

tachypnea Rapid breathing.

thrombocytopenia Deficiency of platelets, structures that aid in the clotting of blood. One of the features of pancytopenia (*see above*).

tight as a drum Phrase used to describe a patient with asthma who, because of the disease, is having difficulty breathing.

tolazoline A drug used in persistence of fetal circulation (PFC—*see above*).

tox screen Short for toxicology screen. A test done on a sample of blood, vomitus, or urine obtained from a patient in whom ingestion of a toxic substance is suspected.

TPN Abbreviation for total parenteral nutrition, in which all the nutritional requirements are supplied via an intravenous route.

transillumination Technique used to "light up" a particular structure. Used to diagnose pneumothorax (*see above*).

traumatic arrest Cessation of cardiac activity caused by a traumatic event, such as an automobile accident.

triage box The place in the emergency room where the charts of the patients waiting to be seen are piled. Patients are triaged according to how sick they are.

trisomy 18 A condition caused by an extra No. 18 chromosome; patients with this disorder are born with multiple anomalies and usually die before their first birthday.

turf Internese for sending a patient to another service.

UA Two meanings: (1) abbreviation for urinalysis, a test performed on urine to see if a UTI (*see below*) is present; (2) abbreviation for umbilical artery, a blood vessel in the umbilical cord that carries blood from fetus back to mother.

URI Abbreviation for upper respiratory infection. Also known as the common cold.

UTI Abbreviation for urinary tract infection.

UV Abbreviation for umbilical vein, blood vessel in the umbilical cord that carries blood from placenta to fetus.

vagitch Internese for nonspecific vaginitis, an inflammation of the vagina.

varicella The virus that causes chicken pox.

V-fib An abnormal cardiac rhythm, the next stage after V-tach (*see below*).

VP shunt A plastic tube inserted into patients with hydrocephalus and that drains excess spinal fluid from the ventricle of the brain to the peritoneal cavity of the abdomen.

V-tach An abnormal cardiac rhythm that, if untreated, can lead to death.